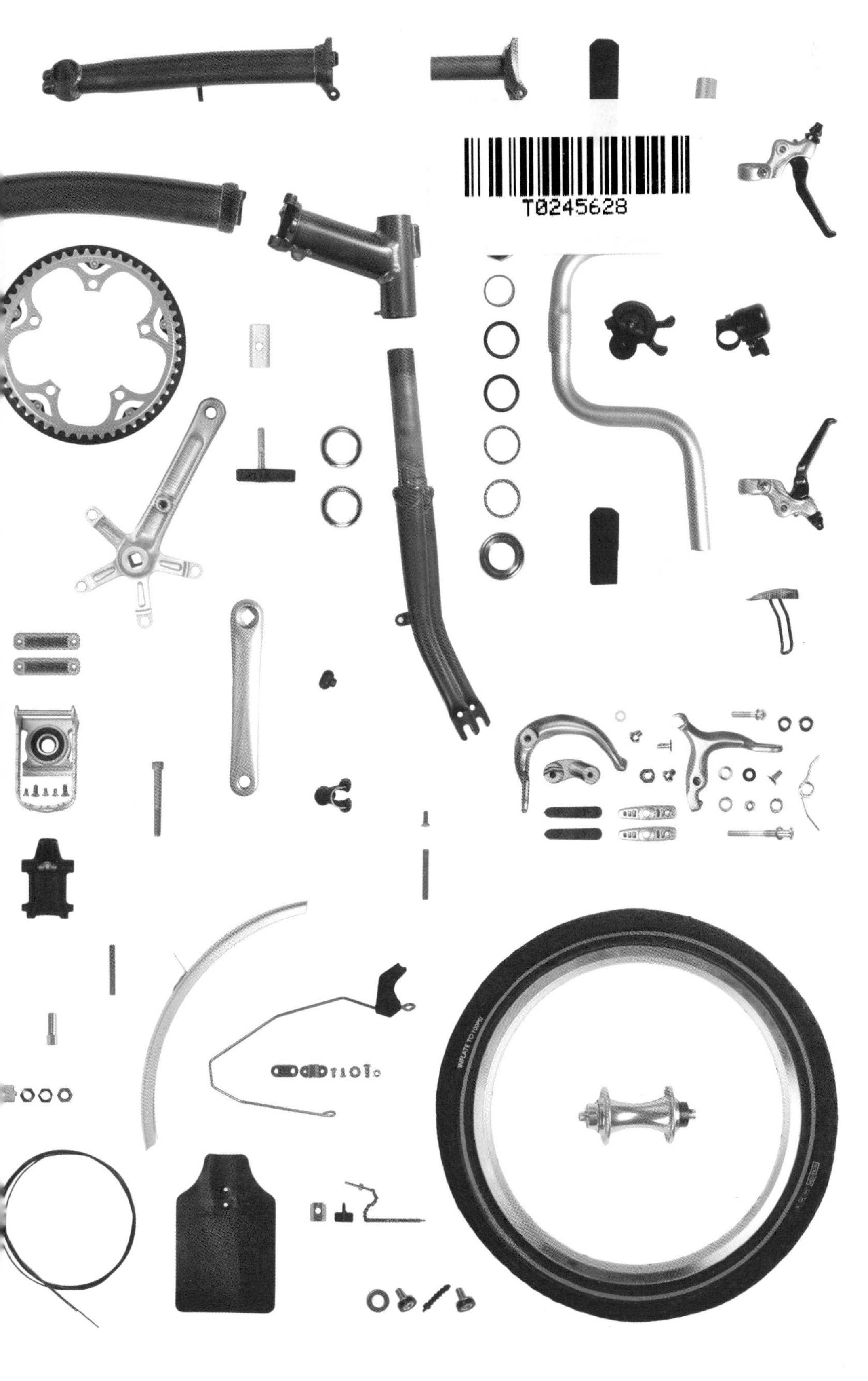
T0245628
INFLATE TO 100PSI

THE BROMPTON

Engineering for change

THE BROMPTON

Engineering for Change

Will Butler-Adams & Dan Davies

NEW YORK

THE BROMPTON: *Engineering for Change*

Originally published in Great Britain by Profile Books, Ltd. First published in North America in revised form by The Experiment, LLC.

The Experiment, LLC
220 East 23rd Street, Suite 600
New York, NY 10010-4658
theexperimentpublishing.com

Library of Congress Cataloging-in-Publication Data available upon request

ISBN 978-1-61519-956-3
Ebook ISBN 978-1-61519-957-0

Cover design by Beth Bugler
Text design by Jade Design

Manufactured in Turkey by Mega Print

First printing November 2022
10 9 8 7 6 5 4 3 2 1

To the B-A team
WB-A

To Tess
DD

CONTENTS

PREFACE

The Brompton bicycle was born out of one man's need. Its inventor, Andrew Ritchie, wanted one for himself, and he also thought that if it solved his own problem it might be useful for others. His resources were scant, but he had time – time to optimise, evolve and simplify his design until he reached an elegant solution:

> The triangular frame holding the rear wheel releases from the seat slide tube and swings under the chainwheel to fold underneath the main frame. A chain tensioner arm swings with the frame to keep the chain taut as it loops back on itself. The angular hinge towards the front of the main frame is undone and folds in parallel with the rear wheel so that the front wheel can be locked onto the chain stay, while still facing forward. The handlebar support hinge is then released, to drop by gravity, locking the handlebars tight in to the front wheel, and finally the seat post is dropped, which locks the whole folded assembly.

If you have ever folded a Brompton, or seen one being folded, you will probably recognise that this is how it works, summarised in just over a hundred words. It's easier to demonstrate than to try to explain it, but once you understand the principles – it's like riding a bike! The Brompton 'just works'. But to deliver 'just works' is not a small thing. It sits on an edifice of engineering detail and thought, and we'll go through the interesting bits in this book.

As well as designing the bicycle, Andrew Ritchie also founded the company, Brompton Bicycle Limited. The company is not necessarily such an elegant or specialised piece of design; it wasn't invented as a unique creation by a design genius. It's a normal business, with human resources, logistics, events, retail and all sorts of other things which don't always lend themselves to engineering solutions. And the company has changed a lot more than the bicycle over time. Since I first joined in 2002, we have grown from 35 staff to 835, and from annual turnover of £2 million to more than £100 million. And we have moved factories twice, to accommodate growth in production from 6,000 Bromptons a year to 95,000.

But the principle that form follows function ripples through all that we do. Everything needs to be lean and to add value. We like clever ideas, we like to do things for ourselves, and we like things that are good value and don't cost too much money. This should be a general principle for any business, whether you're making microchips, food additives or garden furniture. Too often staff in business believe it is their job to do 'stuff', to be busy, but that is not what we want. In fact our world is full of too much stuff, most of it adding no value at all: it is junk. We want more considered thought, time to optimise, so that in every part of our business we deliver something that is clear and simple and moves us a step further forward.

In this book, we will look at how these principles have shaped Brompton, and how Brompton has shaped the bicycle. We're often going to do this by looking at a component in detail, and asking 'How did it get to be that way? Why is it exactly like that, rather than any other shape or material? What sorts of things had to happen and had to be brought together in order to get that part of the Brompton bicycle to be how it is today?' Some of these parts are really important, like the hinges. Some of them might seem slightly trivial, like the little hexagonal boss that holds the plate that holds the chain pusher. In each case, though, because form follows function, explaining the reason for the form starts to tell a whole story about the functions.

But I hope a message will come through that this doesn't have to be complicated. The right thing to do is usually pretty obvious, but is not always the easy thing to do, and may require courage. Keeping hold of your intellectual property in a world of globalised manufacturing is difficult. Finding a way to seamlessly incorporate a 300 Wh battery into a bike, and make it clip on and off intuitively so that the bike can still be carried, isn't straightforward. And planning management succession from a genius inventor who founded the company – yes, that's hard. But none of these things are actually complicated. They require considered thought and honesty, but they don't need endless meetings full of egos, covering whiteboards with vision statements and abstract theories.

We have a vision. Quite simply, we want to change people's lives. The Brompton is a bicycle that disappears when you don't need it and reappears when you do, something you can carry onto a bus or a train, whizz across town or take to explore. It is hardly mind-blowing. It isn't the only folding bike in the world, and it wasn't the first. But, largely because of that hinge, it's a bicycle which people *actually do* carry onto buses and trains, which people *actually do* ride across cities and people *actually do* take on adventures, all over the world.

Even if you only cover four miles a day on a bicycle (two in the morning and two in the evening) rather than sitting on a bus or the Tube or being stuck in a car, that's still 800 miles over the course of a working year. Which works out at about 40,000 calories, which in turn equals about five kilograms of body weight. That won't exactly give you a perfect body, but for anyone past their twenties, the difference between gaining five kilos a year and not is going to add up to something pretty good for your health. On top of which, riding a bike is a blast. It makes you feel good and allows you to declutter your mind, something I for one couldn't live without. All this may sound blindingly obvious, but so often it's not. That is the vision. We know we're doing something that makes life a little bit better, so the more we do, the happier we are.

In order to achieve this vision, though, you have to build a company. One of the biggest traps it's possible to fall into is assuming that since the product is so wonderful, everything is going to work out fine. Probably hundreds of companies with wonderful products go bust every year. If you actually want to change people's lives with a bicycle, you need to let them know that it exists. When they know it exists, they need to be able to find somewhere to buy it, which means you need a distribution chain. Even the task of building enough Bromptons to satisfy demand is not straightforward – for most of the company's life, it's something we've been unable to consistently achieve. It's not hard for an entrepreneur to be tempted to believe that all these problems are difficult, and that therefore they must be a special person even to have tried to solve them. In fact, if you look at problems the right way and have the right people around you, everything is simple, if not always easy.

This, then, is the story of the Brompton bicycle and the company that made it. The story of the people who built the company and how and why we did what we did. It's a story about brazing and titanium and distributor networks and management buyouts and paint inspection. It's told from a particular point of view – mine – and that means there are bound to be some biases and blind spots, but I've done my best to be honest and to consult other people who sometimes remembered things slightly differently. If you're a fan of the Brommie, you'll find a lot in here about how it was developed and some really frighteningly nerdy details about some of the finer technological points. But it's also a story about a company, its working practices and its people, and the business of turning raw steel tubes into a product people want to buy. If you haven't got your Brompton yet, then you might be interested in the story of a company which has, against the odds, and with manufacturing in London, flourished and grown into a global brand. And if you have your own genius idea, I hope you'll find inspiration and a few tips along the way on how you can build a successful business.

I need to say one more thing here before you plunge into the story. It's very simple. Life is short, our moment will come and go, and like our grandparents we too will soon become memories. I reflect on this absolute truth every day. We don't have much time, so the time we do have needs to be used to the full, delivering the greatest positive impact. We live on Planet Earth, a spectacular little blue dot, where we are all connected to each other and to the environment that supports us. But for the last 250 years business has seen its role as delivering profit to shareholders, too often at all costs, resulting in the destruction of societies and our environment all over the world.

Yet those who run businesses, and their management teams, are privileged; they have education, wealth, social stability and healthcare. I believe that for all of us involved in business, privilege comes with responsibility to enrich society and nurture our environment.

Business is not beholden to shareholders. It is the customer who is king. Your business will live or die on its reputation with the customer, and that is something you can never take for granted. The customer is looking for value, meaning a product or service that makes their life better – not because the advert says it will, but because it really does. And to really deliver something delightful you need staff who believe in what they are doing and who genuinely care about how their work affects your customer.

We at Brompton want to change how people live in cities, to bring urban freedom and with it a little more happiness. This is our measure of success, not EBITDA ratios or return on capital employed – although it is fair to say that if we don't look after our EBITDA, our cash, our profitability, we are unlikely to be able to change how people live in cities for much longer. Last but not least, if we get all that right – well, those shareholders who invested their hard-earned savings with us when our founder needed their support to get off the ground will with luck see a half-decent return.

So this book offers a personal perspective on what building a business is all about. And whether you are in the world of business or a Brompton fan, I hope it will be interesting and maybe useful.

Will Butler-Adams

I spent my early career as an analyst in the City of London, trying to make predictions about share prices. In those days, I tended to actively avoid learning too much about the actual businesses involved; there was a huge danger of 'getting married to a recommendation' or 'mistaking the stock for the company'. We spoke to companies roughly once every quarter, when reluctant management teams were dragged out to answer impertinent questions about short-term revenues and profits.

For this reason if no other, it's been a rare treat to be able to spend something just short of eighty hours (a 'nine-day fortnight' in the Brompton working hours system) over the course of a year, talking to the chief executive of a successful manufacturing company about all the things which drive the business, and the human and material reality that lies behind growing the top line at 20 per cent a year. The text is mainly transcripts of Will's explanations and insights, edited and arranged by me so as to tell the overall story and illustrate the key themes. Where necessary, I've added some extra detail to explain exactly what something is, or to fill in facts and figures about Brompton or the bicycle industry in general.

In the nineteenth century an economic journalist called George Dodd wrote a book titled *Days at the Factories: Or, the Manufacturing Industry of Great Britain*. It's an extraordinary project; each of its twenty-five chapters records a different factory, brewery or workshop that Dodd visited, describing in as much detail as he could manage how the raw materials arrive, what kind of work gets done on them and how they leave the factory to be sold. To do a book-length exercise like that on a single company might get a bit mind-numbing, of course, and in many ways leaving out the human stories for a purely technical account would give a false picture. But I hope you'll agree that it can be fascinating sometimes to stop and consider how, in a single product and a medium-sized company, so many things and people have to be organised and brought together and problems solved.

All of the technical engineering and management terms are explained, I promise – although not necessarily on the first occasion on which they appear, if the definition looked like it was going to get in the way of the story.

Dan Davies

A NOTE ON THE BROMPTON FACTORIES

Bromptons have always been made in west London, in progressively larger workshops and factories at a series of locations. Some of the different sites have very similar names, which can be confusing. Early on, Andrew Ritchie had a small workshop beside *Kew Gardens Station*, after which Brompton moved to a slightly bigger space in a railway arch in *Brentford*, where production started in 1988. The next move was to a small factory at *Bollo Lane, Chiswick*, in 1993. In 1998 Brompton moved to larger premises at *Kew Bridge, Brentford*, acquiring additional space nearby ('Unit 19') in 2011, and finally moving to the current 100,000 square-foot factory in *Greenford* in 2015.

CHRONOLOGY

1975 Andrew Ritchie starts designing his folding bicycle.

1976 First prototypes made and Brompton Bicycle Ltd formed.

1981 First production run of Brompton bicycles: 500 made and sold.
The hinge supplier goes bankrupt and Andrew starts looking for finance.

1987 Julian Vereker invests and joins the board.
Brompton awarded 'Best Product' award at the Cyclex trade show.

1988 Production restarts at railway arches in Brentford, west London.

1992 Taiwanese Neobike joint venture agreed.

1993 Brompton moves from railway arches to Chiswick Park factory, west London.

1995 Brompton receives the Queen's Award for Export. The company has 16 employees, making 3,633 bikes in the year, with turnover of £961,000.

1998 Brompton moves to a factory in Kew Bridge, Brentford, west London.

2000 First titanium rear frame prototype made in Korolev, Russia.
Sturmey-Archer goes bankrupt; Sturmey head designer Steve Rickels joins Brompton.
Brompton has 31 employees, making 6,855 bikes, with turnover of £2.05 million.

2002 Will Butler-Adams joins Brompton.
Six-speed Brompton launched.

2003 Fake Bromptons begin to appear at bike shows.

2004 New main frame hinge designed and the Haas Automation CNC system installed.
The company produces 9,888 Bromptons.

2006 Nigel Saffery joins as Lean Manufacturing Manager.
Electric bike project starts.
Superlight Brompton launched, using titanium rear frame and forks.
First Brompton World Championship race in Barcelona, Spain.

2007 Reorganisation of the production line begins.
Brompton now has 71 employees, making 14,401 bicycles, with turnover of £5.28 million.

2008 Management buyout; Will promoted to Managing Director/CEO.
Casting moves from Walsall in the UK to Katowice, Poland.
Brompton M3L added to vehicle collection in the Henry Ford Museum, Detroit, USA.

2009 Lorne Vary joins Brompton as Finance Director.

2011 Unit 19 extension opens to ease pressure on the Brentford factory.
First Brompton Junction shop opened in Kobe, Japan.
First Brompton Bike Hire station (Brompton Dock) opens in Guildford.
First planned launch date of electric Bromptons.

2012 Computerised paint inspection system installed, leading to the Configurator on the Brompton website.
The company has 132 employees, producing 30,902 bikes in the year, with turnover of £16.7 million.

2013 Sixth Brompton Junction shop opens, in Covent Garden, London.
Brompton Fletcher joint venture founded to make titanium frames in the UK.

2014 Brompton buys out Benelux distributor and begins consolidation of distribution network.

2015 Brompton moves from Brentford to Greenford factory.
New investors are brought in. Turnover is £27.49 million, with 42,941 bikes produced by 225 employees.

2016 CHPT3 special edition released.

2017 Electric Bromptons launched.

2018 Brompton Junction USA opens in Greenwich Village, New York.

2020 Brompton Brazing Academy founded.
By now there are 432 employees making 59,062 bicycles a year, with turnover of £57.04 million.

2022 T Line Brompton launched, with full titanium frame.
For the latest full year, turnover was £105.9 million, with 93,542 bicycles produced by 786 employees, despite a pandemic and Brexit.

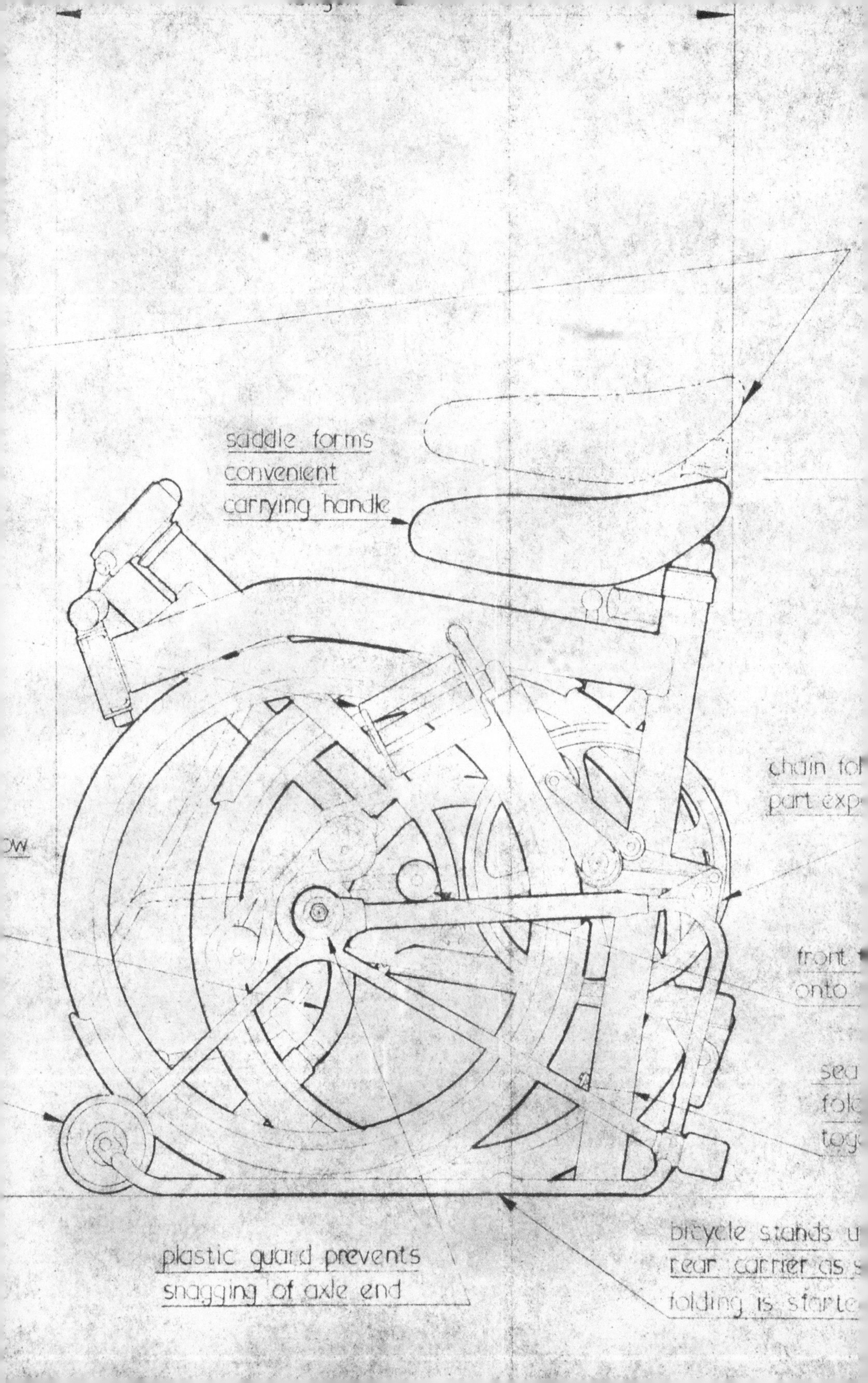
saddle forms
convenient
carrying handle
plastic guard prevents
snagging of axle end

INTRODUCTION: HOW IT ALL STARTED

In 1987, Abdul El-Saidi answered a job advertisement asking for 'torch welders' at what looked like a good wage. He arrived at a small workshop in Brentford, west London, to be greeted by a tall, quietly spoken Englishman. Steel tubes were piled up around the walls alongside boxes of components in a scene reminiscent of a car enthusiast's workshop. The sight wasn't wholly unfamiliar; after learning his trade building power turbines, Abdul had become known as the 'miracle man' to classic car hobbyists, able to recreate vintage parts with almost perfect accuracy. But after a six-hour trial and receiving an almost absent-minded 'Yes, you're good, can you start tomorrow?', he was aware of two things. The Brompton bicycle was an extraordinary product. And this job had nothing to do with torch welding.

In fact, almost everything about the Brompton is unusual. It isn't welded at all – the frame parts are steel tubes held together by molten brass, a process known as 'brazing'. The wheels are only 16 inches in circumference. And at the heart of the thing, there is a fiendishly cleverly designed hinge, which allows the whole bicycle to fold up into a neat package, with all the dirty or oily parts kept away from your clothes as you carry it around. Something like 80 per cent of the parts used are unique to this bike. A product like this isn't – can't be – manufactured by a normal company.

This book is mainly about all the things that I've learned in my time at Brompton Bicycle Limited. I think it's a unique company making a unique product which has the potential to change the world. But conversely, it's a company which works within the same laws of physics and in the same global economy as everyone else, and it's staffed by human beings. So although some things about my experience are uniquely Brompton, many more of them are based on much more widely applicable principles. I'm going to try to explain what I think those principles are and how they're based on straightforward facts, which I'd describe as principles of engineering. This explanation runs through three sections, building on each other to a certain extent, as we see that the same lessons keep reappearing whether you're talking about making bikes, or building a company, or changing the world. But before we start on that, I'll say something about how this company came to be; the invention of the Brompton, and the story of Andrew Ritchie.

Just getting to the stage of having the factory where Abdul did his trial had been a struggle. Andrew Ritchie had invented the bicycle and its central hinge in 1975. Andrew is a tall man who smoked Gauloises cigarettes and wore overalls on the factory floor, supervising everything with a terrifying eye for detail. His story was so extraordinary that it was almost a cliché; you could definitely see Hugh Grant being cast as the young inventor, working part-time on his idea while trying to keep a landscape gardening business going. But that was all true – Andrew really did build prototype parts in his shed, and do the drawings at his kitchen table, looking out of his window at a neoclassical church in Knightsbridge, just down the road from Harrods. The church was the Brompton Oratory, and that's where he got the name.

Andrew was an engineering genius with an unusual vision for the time. During a period when the bicycle industry was almost completely given over to sport and recreational models, he wanted to make a bicycle that would be a useful mode of transport, just as it was in the first half of the twentieth century, but reinvented for the modern city. A bike that could fold up small enough to be taken on public transport and was light enough to be comfortably carried

Brompton's first proper manufacturing facility, in a railway arch in Brentford, west London. Production started here in 1988. Andrew Ritchie is at top left.

would be a kind of 'magic carpet' that you could take from place to place to get you where you needed to be more quickly and comfortably than a car.

But in order to realise that vision, it wasn't enough just to have the idea. He needed to get the thing built. He established Brompton Bicycle Limited with investment from friends. The goal was to make some prototypes and proofs of concept and then sell the plans to an established manufacturer, but none of them was interested (a letter from Raleigh turning him down in 1977 is on page 104). So Andrew realised that as well as being a designer, he would have to become a manufacturer.

This might not have been a natural or comfortable decision to take. Andrew had taken on the project of building a folding bicycle partly because he wanted to own one and thought that existing models hadn't been done right. But also he considered it to be exactly the right scale of project; more or less the largest and most

Julian Vereker, seen here on one of his many early Bromptons, became chairman after saving the company with crucial investment in the late 1980s.

complicated thing that could reasonably be tackled by one person (this was in pre-computer days, when industrial design still meant creating literally hundreds of paper drawings and making wooden models by hand). This meant that he wouldn't have to delegate anything and would be able to keep complete control of the design; everything could be held in his own head. And this felt necessary to him. Compromise and taking easy ways out were pretty much alien to Andrew. They say that you don't often find reasonable people on the tops of high mountains, and something similar is true of anyone who achieves an engineering project on their own.

His early experiences were definitely of the type to reinforce any pre-existing tendencies towards wanting to control everything. He invented an early form of crowdfunding, selling thirty bikes on a 'when built' basis in order to get cash to buy materials and parts, and assembled them himself. About fifty bikes were made this way; it took a while, but by 1981 the concept was proven. Andrew was able to raise money from the first group of investors in Brompton Bicycle and set up production in the Old Power House, near Kew Gardens. After making 500 bikes, however, he lost a supplier – the company producing the hinges went bankrupt. This brought things to an immediate stop. There had been next to no profit on the first 500 Bromptons, but Andrew calculated that with an additional £40,000 of investment he could set up a more efficient process and make a profit while keeping the price the same. At the time, he guessed that it might take up to six months to find a hinge supplier and that the funding could be arranged while he was looking. Instead, over the next four years, every bank he approached for a loan turned him down. We still have their letters.

Early experiences shape companies just as much as they do people, and the lessons that Brompton Bicycle Limited learned in its formative years have influenced the company to a surprising extent, even decades later. Lessons like: 'Money is tight, so improvise and try to come up with creative solutions.' Or: 'The world is full of unreliable people, so keep as much control as you can over everything important.' Or: 'Hardly anybody takes the folding bicycle as seriously as it deserves, so build a community of people who really get it.' And above all: 'Never compromise.' For good or ill, these sort of things shaped the personality of the company.

Me (at left) with Andrew Ritchie at the Brentford factory in 2009.

But there was better to come. Brompton Bicycle Limited didn't always have a lot of luck, but one of the things it had going for it was that people who bought the product loved it. One of them was Julian Vereker, who had invented the Naim range of hi-fidelity amplifiers. Julian had bought six of the first 500 bikes (they were useful on his yacht), and he wanted more. All of his friends seemed to want a folding bike like his, but the initial 500 had all been sold, and he was told he would have to wait six months until production started again. As the delay stretched from six months to four years, he decided to try to get involved himself.

Julian was better placed than many people to see the commercial potential of the Brompton. He made his own product in a factory in Scotland and consequently didn't suffer from the misconception that only big companies were capable of producing things. Eventually he guaranteed an overdraft and became chairman, and at last Andrew was able to raise enough money to restart work in a railway arch in Brentford, within pedalling distance from his flat. That was where Bromptons were being made when Andrew placed his advertisement in the Ealing Job Centre for torch welders whose skills could be adapted to bicycle-frame brazing.

Looking back at the accounts from the days when they were prepared on typewriters, and later dot-matrix printers, the most surprising thing is how little money was needed to make something happen. Julian was a rich man, but there were plenty of other people who could have written a cheque for the £46,000 worth of shares that he bought, even in 1987; an average house in London cost £66,000 that year. By the next year, the company had 'completed its tooling programme and manufactured an initial stock of bicycles', according to the management's statement. There was a bank overdraft of about another £20,000, which then went up to £50,000. All told, it looks like Andrew needed a little less than £100,000 to get Brompton going as a serious manufacturing concern. By 1993, only five years after the investment went in, Brompton had £100,000 in cash on its own balance sheet and had paid back all the bank overdraft. The bike started winning design awards and getting recognition in the press, and had to move to bigger premises in Chiswick, still in west London.

This didn't happen without conflict, however. Julian and Andrew seem to have been cut from similar cloth, and although Julian's expertise was valued (and partly paid for through a consulting agreement alongside his investment), it wasn't always taken kindly. By 2000, the two could no longer work together. Julian sold his shares and left the board; sadly, he died of cancer soon after. But his involvement had an indelible effect on the company; he gave Brompton ambition to grow and to try to change the world, and he insisted on hiring managers and engineers with the capability to build a company that could do so. In 1997, Julian brought in Mike Sear, another audio industry veteran (he had worked for the company that made the Simmons electronic drum kit, which can be heard on practically every disco record from the 1980s). Mike and Greg Smith, Brompton's technical manager, insisted that a bigger factory was needed with room for future growth.

By the end of the century, Brompton had won a Queen's Award for export achievement, but the first attempts to take Brompton to the world spawned a somewhat disastrous venture which still occasionally causes problems to this day. Back in 1992, production had reached a bottleneck at only a few thousand bikes a year. This wasn't enough to keep up with even domestic demand, let alone European exports, and reaching the Asian market was just a pipe dream. Brompton tried to address this by setting up a joint venture with a Taiwanese manufacturer called Eurotai, who would produce licensed copies of the original Brompton. The deal provided for Brompton taking a royalty for ten years on every copy sold, and possibly coming back in the future, when UK production problems had been solved, to sell the real thing. This joint venture company was to be called Neobike. The best thing that can be said about it was that it seemed like a good idea at the time.

This was the early 1990s. Only a few college students had an email account, and there was no computer-aided design software for Windows. Rather than uploading a few megabytes to Neobike, Brompton had to copy and then send by post or courier thousands of physical documents. Not only that, but Andrew had to make copies of the factory jigs and ship them to Taiwan. And then everything had to be explained in telephone calls. Unsurprisingly, this

Andrew Ritchie talking to Yahia El-Sayed Ahmad, one of the brazers at the Chiswick factory.

meant that the copied bikes weren't good. They didn't sell well, and Brompton received hardly any royalties. And of course, all the shortcomings reflected back on the brand; regardless of the quality, cosmetically they looked like Bromptons.

Andrew was bound to take it personally when he was shown a bad piece of workmanship and told that he was associated with it. To rub salt into the wound, the branding chosen by the Neobike management meant that their Taiwan-made 'Bromptons' were the only ones ever sold with a Union Jack on them. Worse still, it emerged years later that Neobike had broken the agreement to return Brompton's designs and tooling once the licensing agreement ended, and fakes started showing up. They still do. It was a formative and bruising experience for the company, and another confirmation that you have to keep control; these things shape a culture.

But even after that, and even after Julian left, the company still wanted to grow. The next chairman was Tim Guinness, an old friend of Andrew's who had gone into finance, and who had a decent idea of what a company with a unique product could be worth. After joining Brompton, Tim went by chance to the 'swan-upping' ceremony at Henley-on-Thames, a classic piece of olde worlde English pageantry in which the members of the livery companies of the City of London assert their ancient rights of ownership over what would otherwise be considered royal birds. Sitting next to Tim on the bus from Henley train station to the reception was a young engineer looking for something interesting to do in life.

That young engineer was me, Will Butler-Adams. At the time, I was studying to begin an MBA, but over the course of a bus ride with Tim, I saw an opportunity to be part of something really intriguing. He invited me to meet Andrew, tour the factory and get a close look at the bike, and I accepted. The visit gave me a clear sense of how far I would have to move out of my comfort zone. From a large corporate setting, governed by processes, structure and hierarchy, I would be going into a fast-growing and slightly chaotic environment, working in a context which, particularly at the beginning, was dominated by the personality and vision of a single inventor. But the bicycle was just so darned good. I wasn't at the time a particularly keen cyclist, but the Brompton is such an attractive piece of engineering that I was drawn to it. I was willing to take a risk.

And that's how I ended up arriving at the Brompton factory in 2002. In an experience surprisingly similar to Abdul's, I took a few weeks doing different jobs and getting to know Andrew before being told that, on balance, I 'would do', with the official title of New Projects Manager. (A marketing guy who joined at the same time wasn't as lucky; he only lasted three weeks.)

My experience at that time was based on having been thrown into running the maintenance of a massive chemical plant at ICI Wilton, on the outskirts of Redcar near Middlesbrough. It was called Melinar 2/4, and it produced tonnes of PET plastic chips to be

One of the first Brompton prototypes, from 1975.

made into Coca-Cola bottles. Two previous planned managers for the factory had turned down the job in quick succession, so there was nobody for them to turn to except a relatively new kid from the graduate training scheme.

It's hard to think of a more polar opposite to the Brompton factory. A chemical works is by its nature not a project that one person can execute on their own, and mistakes can quickly turn into the sort of disaster that appears on the evening news. For this reason, mechanical engineering on this scale is all about control, procedures, teamwork and checking. I had gone there to run a much smaller technical group, and I was thrown into this new role with so little preparation that it was impossible for me to know everything that was going on. I was too young even to bluff. In order to survive, I had to learn to listen. I needed to charm people, gaining the trust of employees who knew what they were doing, delegating responsibility to people down the chain and, crucially, backing people even when I didn't necessarily agree with or understand the course they were taking.

On paper, this might have seemed like exactly the right match for Brompton: complementing the genius founder with a young, affable engineer who was good with people and came from a background of orderly processes and well-arranged plants. Only somewhat later did everyone begin to realise that 'Always wants to delegate' and 'Wants to keep control' are not necessarily styles that complement one another; they're very deep-rooted philosophical differences. We were all going to go through many more learning experiences over the coming two decades.

PART 1

MAKING BIKES

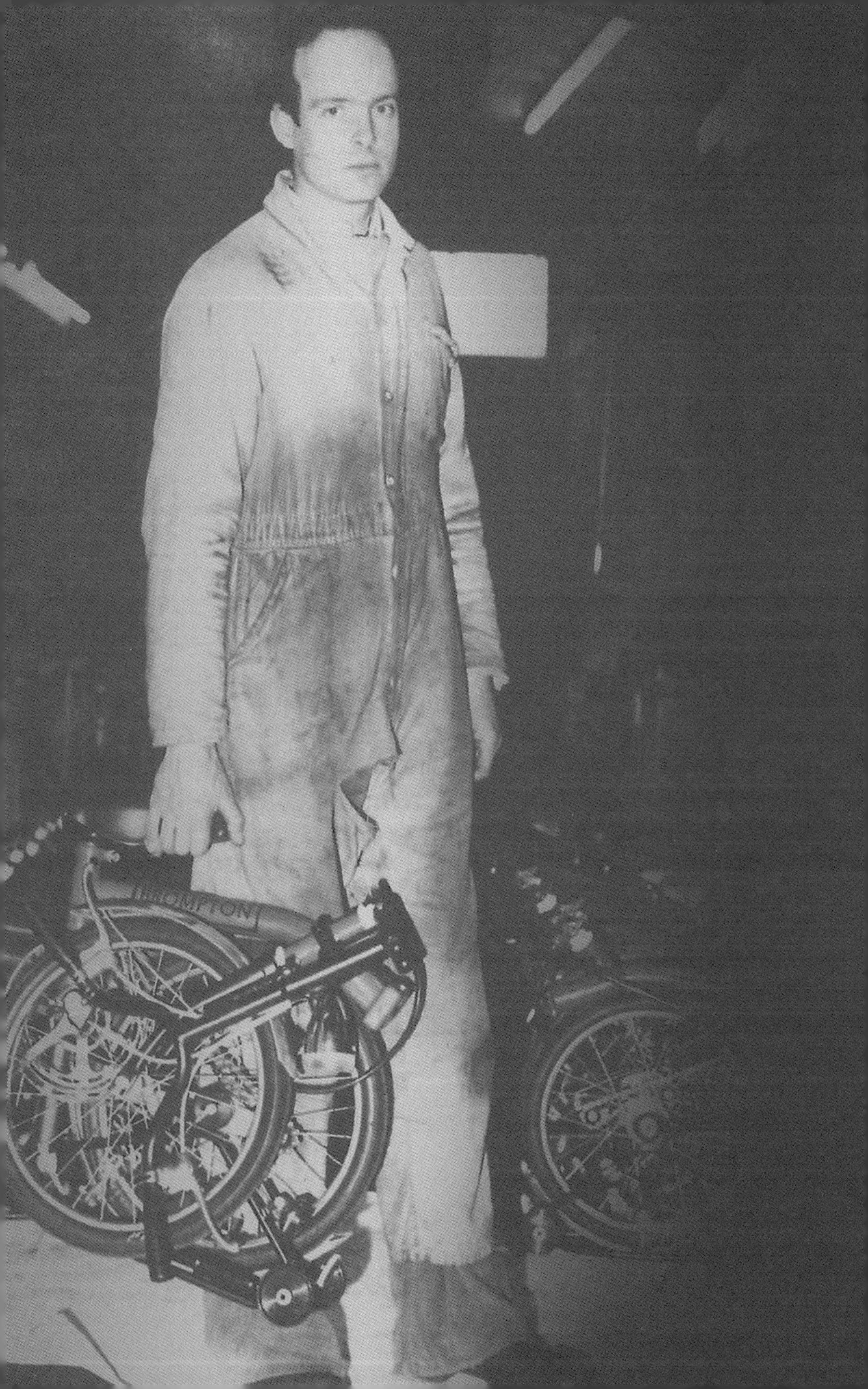

Most people don't see things being manufactured these days. If a new startup business has the word 'factory' in its address, it usually just means that it's based in a nineteenth-century brick building with a big hall that's been converted into offices. We tend to push our noisy, smelly industrial processes to the edges of cities or out into business parks. Or even better, send them overseas and let other people handle the manual labour, keeping the valuable, profitable and clean jobs in brightly lit offices in the nice part of town. It's not an economic model I agree with, and I'll discuss that later. But for now, I need to explain why, before talking about culture, finance, design, intellectual property or any of the other daily activities of a twenty-first century manager, I'm going to spend some time discussing in detail things I learned about the manufacture of the Brompton bike during two decades in a factory.

If you look at most things closely enough, you find out something fascinating about them. This is particularly true of an invention like the Brompton, where the central mechanism is clever and practically every major part represents the outcome of many different engineering problems that needed to be solved. Factories throw up tricky technological questions, and they're full of larger-than-life people, and so if there's something to learn about general industrial principles, it's likely that there's a story from the floor which illustrates it better than a diagram or a page of abstract reasoning. But there's a more important reason to start this way, which is that, as well as literally pushing manufacturing industry to the edge of town, modern society seems to have pushed it to the intellectual outskirts as well. Galileo was a lens grinder, Newton made his own equipment, and Leonardo was always manufacturing new devices. But these days, it seems almost funny to think you could get a deep insight into anything important just by thinking about making things. And that's a shame.

Over the couple of hundred years since the Industrial Revolution, people involved in running factories have faced a lot of recurring problems, and they've come up with general solutions to a lot of them. This is a body of knowledge – it's not really an academic discipline, although some of it is taught in universities as part of engineering,

The Brompton team soon after moving to new premises at Bollo Lane, Chiswick, in 1993. Andrew Ritchie is at the far left.

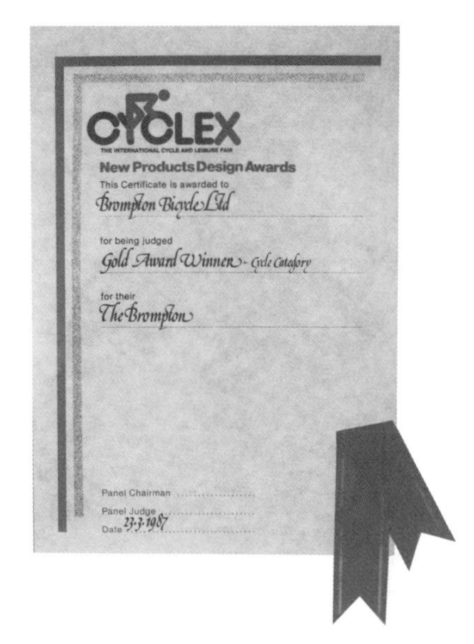

Regardless of manufacturing difficulties, Bromptons won industry awards from the very start.

and some of it is taught in business schools. In many ways, it's a pity that nobody picked up all these solutions and systems and turned them into something which could be taught to undergraduates or even high-school students. Because they really are *general* methods of problem-solving; there's no reason why they should be thought of as particularly to do with manufacturing or business. Everyone has to deal with complexity, with systems that interact, with skills that are difficult to learn and, most of all, with information which arrives more quickly than an individual can process it. These are the things which you really have to manage in order to turn a truckload of steel tubes into a hundred folding bicycles; the physical and chemical properties of the processes involved can be looked up in a book.

The great thing about manufacturing is that everything is visible and tangible. In other industries, people talk in abstract concepts, sometimes barely measurable. Performance measures have to be agreed upon and may be debated over and over again. The connections between different divisions and operations are shown

Andrew Ritchie being presented with the Queen's Award for Export at the Chiswick factory in 1995.

on an organisation chart – or often they aren't, because they may be obscure and implicit. This tends to mean that business philosophies and management principles are equally abstract. But in a factory, the chain of cause and effect is right there, laid out in front of you. There's still some room for debate about what the output was, or what inputs were required to make it, but there's much more objective reality in the room. And the connections, workflow and bottlenecks manifest themselves quickly and obviously – physical objects tend to pile up or run out, and there's no concealing that.

This means that manufacturing is a good place to learn lessons that are applicable to all sorts of businesses. Everyone has bottlenecks, quality-control issues and problems with introducing new processes. And all businesses change as they grow – the things which worked when you were building the company never work in exactly the same way when it grows. Everyone has interpersonal conflicts, and every manager has to deal with the fact that one person can't be everywhere all the time. As I learned about making bicycles and went from being the engineering manager to managing director to CEO, I made important mistakes, but I hope I made them in *clear* ways and learned from them. So this first section of the book deals with physical processes and how they illustrate some important strategies for solving problems. You can learn a lot about the world just by looking at folding bicycles, if you look carefully enough.

CHSSET
Check that spacing at each end falls within the ranges below.
Chainstay forward-end
Chainstay rear-end

1 UNDERSTANDING THE JIGS

> *In engineering contexts, 'tolerance' is a measure of the accuracy of a measurement or the permitted variation in a dimension of a component. Typically in the bicycle manufacturing industry, tolerances of around 2 mm are normal. The Brompton bicycle is made to tolerances of 0.01 mm for individual measurements and 0.2 mm overall.*

'A waste of money … completely unnecessary … exceeded your authority … ridiculous!' It was one of the first capital investments that I had ever made at Brompton. It cost £21. It was a digital weighing scale.

Everyone has arguments at work, but this one has stayed with me for twenty years. It happened only shortly after I joined the company, and it was the first time that Andrew Ritchie really lost his temper at me. In many ways, it was a trivial argument, even a ridiculous one – a weighing scale is just a scale to weigh parts, and if you want to save weight in a bicycle you need to know what something weighs, not to within a few grams but to within half a gram. Andrew's reason for objecting was that he felt the small red plastic set of scales already in the factory, borrowed from his kitchen, was perfectly adequate to the task.

This row sticks in my mind because it was so out of character. Even today, when people ask questions about our manufacturing process, however technical the answer, it always seems to begin with the words 'Well, the thing you have to understand is that Andrew is a perfectionist'. Making do and muddling through when it came to measuring was absolutely not like him, and it probably indicates that (as anyone who's been in a relationship of any kind will tell you) arguments are never really about the thing they seem to be about. They're always about the bigger issues, the icebergs under the surface. The weighing scales row happened because someone was going around doing things in the factory without telling Andrew, and making decisions of their own rather than acting as a human solution to the fact that he couldn't be everywhere. People talk about 'founder's syndrome' in companies, and at this point our founder had reached the limits of his tolerance.

In engineering contexts, 'tolerance' has a specific meaning, referring to the accuracy of measurements and the variability of components. It's a concept that comes into use because you design things as abstractions – perfect circles, exact weights and precise measurements – but then have to make them out of actual physical

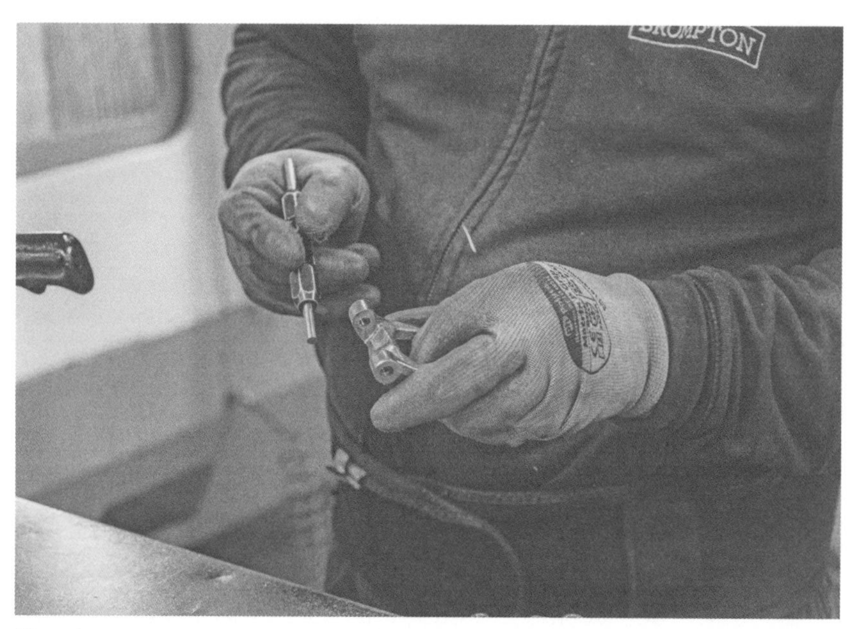

Checking the tolerance of a hinge spindle using a 'go/no-go' gauge.

objects which have physical limitations affecting the precision of the end result. Depending on the application, you might need a very small margin of error indeed, or you might be able to accept a bit more variation in order to keep costs down. That's why we use the same word which in ordinary language means the ability of people to get along together – just as tolerance between people involves making compromises with one another, tolerance in engineering is a measure of the compromises you're prepared to make between your perfect design and messy reality.

A folding bike is, fundamentally, a massive compromise by its very nature. Placing hinges in a bicycle frame is exactly the wrong thing to do in terms of the properties you want a bicycle to have. Every imperfection is a potential source of movement, and even tiny wobbles are noticeable to the rider. So, having made that initial design compromise, your scope for making further compromises is greatly limited. And tolerances tend to add up; a fraction of a millimetre in one component can be magnified by a similar amount in another one, and before long you can end up being three or four millimetres out at the other end of a long tube. Three or four millimetres is definitely enough to affect the ride quality; if the floor underneath you right now moved by half a centimetre it would feel as if there had been a small earthquake.

This was a key problem which affected nearly all folding bikes before the Brompton, and their manufacturers tended to try to solve the wobbliness issue by doubling up on weight. One of Andrew Ritchie's key insights was that the Brompton would have to be manufactured to much greater precision than non-folding bikes; in the present-day factory at Greenford in west London things are generally calibrated to plus or minus 0.2 mm tolerance. This means that the final tolerance in the overall wheel alignment, once all the frame parts are put together, is no more than 2 mm. A factory making a non-folding bike, of course, only has one frame part to worry about, so they can work to 2 mm tolerances. If we did the same, the wheels could be out by plus or minus two centimetres.

It's worth thinking about what this sort of accuracy means in practical terms. Try holding your finger and thumb consistently half a millimetre apart for a whole minute. You'll find you can't do it; your muscles start to tremble. At this level of precision, you have to take into account the rigidity of your tools themselves. Other industries, like automotive and aerospace, operate to even finer tolerances and have to take into account the potential effect of a few degrees' difference in temperature on the expansion of metal. Brompton is probably working at the limit of what can be achieved with a mostly handmade product.

But how do you get that sort of precision? In a manufacturing company at least, you start to build jigs.

'Jig' is a gloriously old-fashioned word that isn't quite part of everyday language any more, now that fewer people work in factories. But a good jig is a thing of beauty. The principle is simple – a movable piece of equipment holds a component or a tool, or both. It controls the motion so that the path taken by the tool over the component is exactly the same each time. Effectively the jig is a kinetic sculpture of the finished piece. The range of movement of the jig corresponds to some part of the shape, or at least to the curve followed in space and time by some operation that has to be performed. To a very large extent a manufacturing process *is* its jigs; to understand the factory is to understand its jigging, and vice versa.

The concept of a jig generalises into a principle that goes well beyond machine tools. So it's important to really understand it before we go on. If you're not an engineer or a woodworking hobbyist, say, it might help if I give a really prosaic example.

I once found myself needing to rescue someone from their flat-pack furniture nightmare by drilling six new dowel holes into a piece of wood. The holes had to be exactly perpendicular, and they definitely had to be of a consistent depth so that they would hold the dowel and not come out through the other side. So I needed to improvise a drill jig. An 8 mm drill bit fits snugly through the centre of an ordinary cotton reel. If you slide the reel onto the bit before clamping it into your drill, not only will it hold the bit perfectly perpendicular as you move down, it will also limit the depth of the hole. The cotton reel, in this case, is acting exactly as a jig does in a factory – it's constraining the movement of the tool in order to ensure that the desired operation is carried out consistently.

The jigs used in the Brompton factory are more complicated than this because they have to take into account a greater number of operations. They are also significantly more precise, because they've been custom-made rather than bodged from things lying around the house, but the principle is the same. The jigs make sure that each component assembly is put together right, by taking away the freedom to make the wrong moves. And it's possible to use the same principle to verify the process. A 'go/no-go' gauge is like the jig version of a check. If the part will pass through one gap in the gauge, but not through a gap of a different size and shape (the concept is pretty similar to the sorting boxes that babies play with to learn about shapes), then it can go on to the next stage of the manufacturing process.

In Japan, the go/no-go concept has been extended to a whole system of practices. Anything which physically prevents a mistake from being made, or automatically generates an alert that something has gone wrong, is a *poka yoke*, invariably pronounced 'pokey-yokey' by British engineers. (*Poka* in Japanese is a mistake, and *yoke* is to avoid something – it is the equivalent of 'fool-proofing', but in a language that's a little more polite about calling people fools.) The protruding top pin on a British electric plug, which won't let you put the live and neutral pins in until there is an earth connection, is in the spirit of *poka yoke*, and so is the plexiglass cover that (usually) stops a fighter pilot from accidentally pulling the handle to fire the ejector seat. The idea here is that the information needed to solve the problem is generated as part of the overall process.

Andrew was fantastic at designing jigs; as our former manufacturing manager Nigel Saffery says, some of them were works of art. They had moving parts in multiple axes, magnifying glasses attached to help the worker and even tables engraved into them to remind you which bolts to use where. (For his part, Andrew once wrote in a memo that Nigel was 'probably the only person who understands all of the factory's jigging', a compliment of the highest kind.)

This sort of thing is the basis of mass production. An eighteenth-century cabinet maker might have crafted everything with a set of chisels and made every measurement individually, but even then, the lower-order craftsman who made chair-legs for him would have used some sort of jig to make sure they came out looking identical. Losing our jigs would therefore have been a disaster – the kind of setback that destroys companies. It's hard to believe that in 2004 we actually semi-accidentally inflicted this on ourselves.

It happened like this. On some bicycles the main hinge was failing in the field after around ten years of heavy use, due to metal fatigue. This wasn't good enough for us. We had traced the cause of the fault; the steel around a joint was inevitably weakened by the heat used in joining the two pieces of metal. In order to improve this, Andrew had redesigned the hinge, making the joint a little bit more complicated, with a piece extending into the body tube to add strength, rather than butting two flat surfaces together.

The new design was a lovely piece of engineering. It had three times the fatigue life of the previous design and no more weight, while also giving us a chance to extend the frame by three centimetres, which Andrew had wanted to do for some time. But the sockets into which the frame tubes were inserted needed to be an exact fit. And that meant exact, because the main frame hinge is attached to the longest tube in the frame, and a small inaccuracy at one end of a long tube results in a significant displacement at the other end. An error of only a fraction of a millimetre at the hinge would affect the ride quality and the life of the bike.

Perhaps surprisingly, the first stage in making this hinge employs the same sand-moulding method that the Romans used 2,000 years ago. In our case, an aluminium pattern is pressed into fine sand and removed before molten iron is poured into the depression. The

The Brompton factory's Haas CNC machine, a high-precision automated manufacturing tool used for the redesigned main frame hinge.

finishing of the cast parts for the redesigned hinge – the process of taking the pieces of iron produced by the casting process and machining them down to smooth parts of exactly the right dimensions – was going to be beyond the capacity of a human machinist, however. It would require computer control.

In the metal machining industry, the relevant piece of equipment is called a CNC machine. The initials stand for 'computer numeric control'. One might think that 'computer numeric' is a tautology, but in fact the electronic computer took over from mechanically controlled machine tools that had been in use since the nineteenth century. Even today, some of the most sophisticated CNC machines are sometimes referred to as 'Swiss lathes', after the complicated and extremely flexible tools used by watchmakers. A modern CNC machine is a very complex, very tricky piece of technology.

Andrew characteristically decided to make his own CNC machine. I didn't really think this made sense. He had a number of views on the design of commercially available CNC machines and felt he could improve on them, but although his ideas may have been fundamentally sound, the project was just too big for one person, however talented. And since Andrew's time and expertise were two of Brompton's scarcest resources, I couldn't see the sense in doing something which diverted him from the business of designing bicycles.

By the time the new hinge was under development, Andrew's bespoke CNC system was already making the handlebar support hinges, but it wasn't going to be up to the more complicated work on the main hinge. It wasn't flexible enough to handle the subtle differences required for the similar but different main frame hinge. Ironically, it was also too flexible in the literal sense – despite having been built with huge steel I-beams, its structure was not quite rigid enough. At the kind of tolerances we were talking about for this part, even the tiny amount of give in a girder is too much. High-end CNC machines are cast out of a single piece of metal and are often so

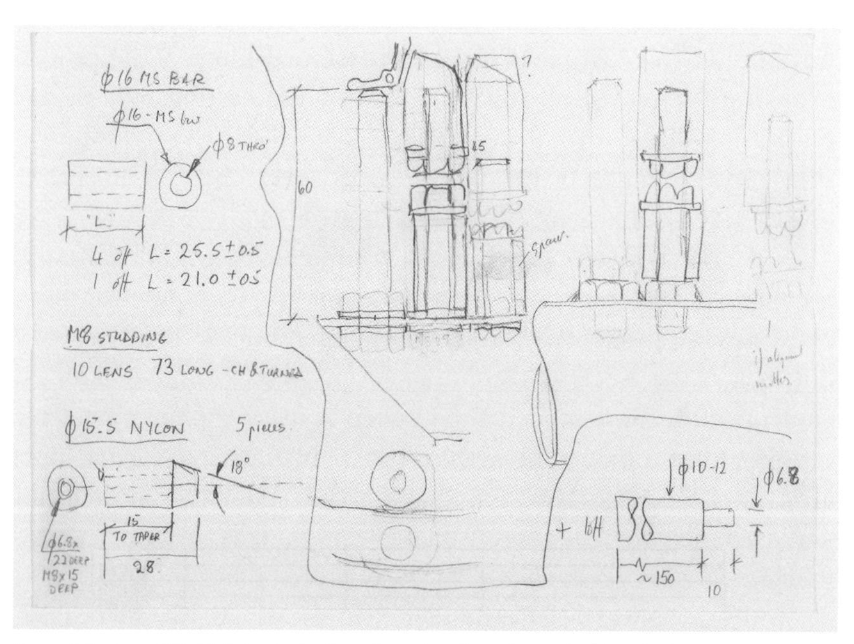

One of Andrew Ritchie's technical tooling sketches, from the days before computer-aided design.

heavy that you need to reinforce your floor in order to install them. However much ingenuity we applied, it was not going to overcome the fundamental issue of materials science. We were going to have to find the money and buy a machine.

A frank and spirited discussion ensued. A machine was going to cost about £35,000 (considerably more than any set of scales). Andrew was still not impressed by the manufacturers of CNC machines. But we needed one there and then. We couldn't make enough bikes to meet orders, so there wasn't time to reinvent the CNC. Even if the rigidity problem of the I-beams could have been solved, the new hinge couldn't wait for Andrew to design a masterpiece. In the end, what brought him round was the question of spare parts. If a bespoke machine breaks down, you have to fix it yourself and also manage without while you're fixing it. But a commercial machine comes with warranty support and a team on call 24/7 with a full stock of spare parts to get it back up and running again in no time.

The California-based US company Haas Automation (owned by Gene Haas, sponsor of the eponymous Formula 1 and NASCAR racing teams) got the contract. There's really no such thing as an entirely off-the-shelf CNC system, and the Brompton hinge pushed the envelope for what the Haas VF-1, their flagship model of the 2000s, could do. Bought as standard, it was a three-axis system, which could move a tool head across an object in three dimensions. But the machining of the new hinge would require rotation as well as forward and back movement. Theoretically it would have been possible to move the part, part way through the machining process, but every time you remove a part and reorient it to allow the tool to access, you introduce variation – you can't put it back in exactly the same place, or at least not to within the tolerances that our hinge demanded. We worked with Pat Fenn, the head of the Haas UK distribution company, to install a turntable and a driven drill and tap head, effectively turning the CNC machine into a five-axis system. More challenging, though, was the design of the holding fixture. The hinges needed to be securely held in exactly the same place every time, with enough room to ensure that all the cutting tools had access. Also, the hinges had to be secure so they would not move when being machined, but if they were held too tightly they would become distorted while in the fixture, and when removed they would not be accurate.

Delivering a system that met Andrew's requirements was not easy. The Haas technical centre in Leicester was put through its paces, and it still advertises us on its website, celebrating that its technicians finally won our approval. At last Brompton was ready to start producing the new hinges. There was just the small matter of the braze jigs.

Obviously, it wouldn't be possible to use the same jigs for the new hinges that had previously been used on the old ones; the hinges are different, and the frame was now three centimetres longer. But instead of creating new jigs for the new design, Andrew decided that we should modify all the existing jigs.

With the benefit of hindsight, this was crazy. It was in keeping with the spirit of the company to strive for economy, and by modifying we certainly saved a few thousand pounds – but at the same time we destroyed our production jigs! There was no going back; from the moment the tooling change was made, it would no longer be possible to produce bicycle frames until there was a supply of the new style hinges. This meant that an unwanted race against time had suddenly begun. Every bike we made was now depleting our stock of finished frames.

And although the modified Haas CNC machine was a good tool, it was not yet ready to go into full use. The full complexity of the system that had been put together for Brompton was only just becoming apparent. The tools had to be calibrated, and the holding fixtures placed in exactly the right place relative to the cutting heads. But it seemed that every change that was made affected something else in the system – not by much, but even fractions of a millimetre mattered. It was an atrociously fiddly task, like tuning an oversized grand piano. And the trouble with a task like this is that it isn't incremental; you never know how close you are to finishing until the moment you get it right. All the while, we were running out of frames to make bikes.

Eventually, and in the nick of time, we managed to get the machine tuned, and socket hinges started coming out to within the necessary tolerances. Production never needed to stop, and the new design worked. But things had been cut far too close. The cost saving from modifying the old jigs rather than building new ones could not have been more than the sale price of two dozen bikes; the downside risk of losing weeks of production was totally disproportionate. This at a time when – as has been the case for much of Brompton's history – the major problem faced by the company was long waiting lists and inability to make enough bicycles to satisfy demand.

Looking back, it's obvious that we'd thought this through almost exactly the wrong way round. We were solving the problems in front

The Haas machines needed to be precisely set up and calibrated. Torsten Steinebrunner was one of the engineers responsible.

of us, but we had started from the new hinge that we were excited about. The new hinge needed a new CNC machine, so we got one of those. And attaching it to the bike would require new jigs, so we looked ahead and adapted them. Every individual step of the way, we had done things as efficiently as we knew how. It was a perfect example of the engineering mindset.

Except that it wasn't; it was an example of a bad engineering mindset. A correct approach would firstly have focused on the thing that the rest of the system exists to produce – a constant flow of bicycles out of the factory – and worked around that. Secondly, it would have noted that fine tolerances are difficult, and that tuning the CNC is the sort of task where it's hard to know how long it's going to take until you start doing it. (One thing that engineering training does help you with is recognising that some operations have this property – they're the ones where 'nothing's finished until everything's finished' and changing one thing affects another. But seeing what's right in front of you is often surprisingly difficult.)

If we had put together those two fairly simple ideas, we would obviously have spent the extra money and built new jigs. We might even have planned how much stock of old-style frames to keep. But the lesson isn't the simple technical one of 'don't adapt your production jigs'. It's the much more general point that when you start to make a plan or solve a problem, you need to take it seriously – to have respect for the problem.

I do find that this is sometimes difficult to get across when we recruit people. When I was at the Melinar chemical factory, it was psychologically a lot easier for everyone there to have a shared understanding of the importance of quality, because the consequences of failure were potentially and obviously disastrous. But with a bicycle, there's a tendency to treat it like a toy, with toy problems. This is bad. If your bike fails, you can end up on the tarmac with no front teeth and a broken pelvis. If you are really unlucky you may even be hit by a car or a truck. We have nearly a million customers out there in the world, and we need to look after all of them for the life of their bike.

Treating a problem with respect means understanding that the thing you are doing will have an effect on the final product which you are sending out of the factory to sell to customers. If it doesn't have an effect, why are you doing it? And that means that your analysis needs to start at that end: the thing that generates the value. You can then trace the chain of cause and effect back to the thing you want to do, the new machine or shiny redesigned component. Like so much, it's basically simple, but the challenge is having the strength of mind to do it consistently.

2
THE ART OF BRAZING

Brompton bicycle frames are brazed at a temperature of 900°C. This is significantly lower than the temperatures used in steel welding, and means that there is much less heat stress on the metal. Every brazer has their own mark, which is left on every joint they make, to allow their work to be traced.

Rebecca Francis joined Brompton at the age of 19. Her surname was Summers then, and after being made redundant from a job in a hairdressing salon, she went into an entry-level role on the 'pre-pre assembly' team, putting together some of the smallest components. She fitted in well, and she was hard-working and willing to listen, so she stuck around. After a while, she noticed that the best-paid workers were all in the brazing department, and that none of them were women. We have a rule that anyone who wants to can try out to be a brazer, and that's what Rebecca did. Ten years later, she's one of the best brazers in the world. That's because she's one of the best at Brompton – and in this particular craft, Brompton is unarguably the top of the tree.

The joints in the frame of a Brompton are brazed, not welded. In brazing, the steel is never melted; instead it's held together by a molten 'filler' metal. (Originally, the metal used for this purpose was brass, hence 'brazing'.) Because this method causes less heat damage to the tubes and leaves them stronger than in welding, brazing allows us to use thinner tubes and save weight. For this reason if no other, brazing is intrinsic to the Brompton; it's right at the core of what we do. If you were to visit the factory, you would see that the staff recreation areas are full of things that have been brazed together, like our coffee bar. The golden brass line between two surfaces of steel is a visual signature for the company, as well as being the best way to join steel tubes. This has always meant that anything that happens in that department is going to have a chain of consequences that ripple throughout the whole business.

As Abdul El-Saidi found out on his first day in the railway arch in Brentford, brazing is more difficult than welding. The essence of brazing is surface tension, the same physical principle by which water soaks into a paper towel or gets pulled into the roots of a tree. The filler metal melts and then flows into the microscopic cracks between the two surfaces, bonding them. But the metal has to be able to flow, so the surfaces have to be clean, and the gap between them has to be just exactly tight enough.

The parts are held in a braze jig, which holds the surfaces against each other with precisely the right amount of contact. The type of contact, and hence the form of the jig, depends on the type of joint. For the bottom bracket, for example, the joint is 'butt brazed', meaning that the pieces of metal to be joined are pressed together ('abutted'). The redesigned main hinge which we talked about in the last chapter is 'fillet brazed' or 'socket brazed', two terms that mean more or less the same, depending on whether you are focused on the hole (the socket) made in one piece of metal, or the extended piece on the other (the fillet) which fits into it.

There is also the matter of flux. Etymologically, flux in welding applications is 'that which makes the metal flow'. Chemically, its main purpose is to get rid of any layer of oxide that has formed on the surfaces to be brazed, because the oxide layer effectively 'waterproofs' the underlying metal, stopping the braze from bonding to it. Getting the amount of flux just right is vital; too little and the braze won't flow properly; too much and the excess flux residue will leave a 'glassy' deposit that compromises the strength of the

Bottom bracket joints showing a fine example of 'linear' flow brazing.

joint. Small brazing shops might use flux powder or paste, and then clean it off just before starting to heat the metal, but in the set-up used at Brompton, the flux is introduced by a 'gas fluxer' – acetylene is allowed to bubble through liquid flux before being combined with oxygen at the torch head. You can identify a gas flux system by sight; a neutral oxygen-acetylene torch would burn white, but the Brompton brazing system has a green flame, and an experienced brazer can tell from the precise shade of green whether the fluxer is correctly adjusted.

The heat has to be applied to the metal being bonded, not to the brazing wire. Beginners might make the mistake of melting the wire and dripping molten braze onto the joint. This is called a 'cold joint', and it won't hold. The name summarises the problem: in a cold joint, the inside of the joint won't be hot enough to keep the braze liquid as it flows into the gap, so the braze metal won't be pulled far enough into the joint. It will just flow over the outside, leaving a weak join.

If you get everything right, then a good brazed joint is 'linear' – the molten braze looks like a smooth flowing stream that's been suddenly frozen. It doesn't have bumps or smears or lateral wiggles; it reflects the smooth and co-ordinated movement which produced it. It isn't glassy, indicating that too much flux was used, and it isn't discoloured, indicating that the temperature was too high. The metal bulges with 'ripples' or 'stitches' which actually strengthen the joint as well as ornamenting it. This is the standard that we refer to as 'raw lacquer quality' – frames on which all the joints are aesthetically pleasing enough to be given a clear paint job which exposes them to view. Just below that in quality terms are joints that are not quite linear or where the stitches are uneven. That is the minimum standard for a Brompton with opaque paint. And obviously nobody in their right mind would let a weak or cold joint out of the door; almost all the brazed joints on the frame are safety-critical.

We have machines which handle the brazing of the hinges, but there is no way that they could tackle the frames. Automatic brazing machines are, if anything, even more temperamental and difficult to calibrate than a CNC system is. As well as needing precision in the positioning of the part, they need their gas mixtures to be just right, which is not always easy to achieve when the nozzles are constantly clogging up with flux. The first autobrazer that the factory had was made to a design of Andrew's, and there was only one employee who could operate it; when he left the company, I had to do this job myself for three months. As technology has advanced, and we have ordered new machines, we've been able to stabilise things somewhat; the second machine that we added has thermographic cameras, which allow the machine to simulate something like the feedback loop of an experienced brazer watching the tip of their torch flame. The third-generation version even has an air chiller built into it, so that it doesn't need to use water, which isn't so good for steel, to cool the parts back down to handling temperature.

But each of these machines is a huge and expensive piece of engineering, and setting them up for a job is still a serious operation.

The frames have multiple braze joints of both socket and butt types, and they need to be turned around and reset in different jigs for each braze to be made. That means that hand building is the only realistic option.

Producing high-quality hand-brazed joints is not something that just anyone can do. As well as a steady hand and an ability to think in three dimensions, it requires an unusual degree of calm and concentration. You need to shift your point of focus to the tip of the flame, but also split your awareness to pay attention to the tip of the wire, guiding the metal to the correct point and keeping it moving. Granted, it's not brain surgery – but neither is it wholly different from the manual activities of neurosurgeons. In order to braze well, you need to be in the same sort of 'zone' of active concentration as a sportsperson. Keeping that level of concentration is tiring, and not everyone finds it pleasant. Small mistakes can bring you out of the zone, and they can be just as frustrating as a sliced golf swing. It's a skill that not everyone is capable of acquiring, and even those who manage it may not become good enough to work at a reasonable speed.

In other words, it's a difficult job, and a good brazed joint is a piece of craftsmanship. Every brazed part at the Brompton factory is stamped with the initials of the person who brazed it, partly to track quality control, but just as much to recognise that it's something unique, produced by a human being. This means that the availability of skilled brazers is itself a bottleneck. When I joined, there were six full-time brazers in the factory. That meant that if two people were on holiday and another one was sick, production fell by 50 per cent. It was not uncommon to have to reallocate people from the assembly line to other tasks because the stock of frames had been used and there was nothing for them to assemble.

Brazing is close to being a lost art. You can't put an advertisement in the newspaper or ring up an employment agency and hope to recruit a dozen brazers. All you could do, as Andrew did in the early days, was try to recruit torch welders, and hope that out of every

Master brazer Abdul El-Saidi at the Brentford factory in 2010.

ten or twenty that applied for the job there might be one who could make the switch. In order to expand, Brompton had to start training people ourselves.

We were lucky in this regard to have Abdul El-Saidi, a big and generous man with a gift for molten metal. Abdul had learned welding in a car workshop, then perfected the art on the site of a huge turbine project in Lebanon which was unfortunately cut short by the 1979 war. His teacher there was a German master craftsman, someone he occasionally describes as being like a 'professor of welding', who organised competitions between welders from all over the world at the site. Ever since the early days with Andrew, Abdul had been the one showing people how to do things and improving the processes – he even invented a way to improve the corrosion resistance of the rear frame by doing without two little holes which had originally been made to prevent a vacuum developing inside the tube as it cooled. He was a natural choice as training manager.

The biggest problem we had with Abdul was to try to stop him from taking on too much – he would always be trying to manage the brazing shop, train new brazers and at the same time braze more frames than anyone else on the shop floor. We tried to get local universities and colleges involved in our training, but while they could help with background knowledge of the chemistry and physics involved, the instructors tended to be from a welding background themselves, rather than being able to pass on the specific skills we needed. On some occasions, people have had a hard time believing that the work we're asking them to reproduce was done by human beings. The trouble is that, in the UK at least, the only places where steel tube brazing by hand is routinely done at a high level are either Brompton or a small number of custom frame manufacturers, many of whom are themselves former Brompton employees. There's nobody else to share our training and development costs with.

It is possible to turn this problem into a strength, though. If we have to train our own brazers, we can train them to be exactly the right kind of brazers for Brompton bicycles; nearly all the learning in our training facility is done on specific tasks which we need, and someone's progress through the curriculum is marked by which joints on a Brompton frame they're capable of being certified for. One of the benefits of doing things for yourself rather than buying them in is that you can get things done exactly the way that you want them, rather than compromising with all of your suppliers' other customers. This is true of training just as much as it is of machines and components.

The brazers have a special skill; there aren't very many of them; and they are completely vital to the overall production, because the line as a whole can't move faster than the output of frames. In bicycle racing, your team is judged according to the position of its fastest member, and team tactics are all based on this fact; everyone on the team works to get one of the leaders across the line as quickly as possible, often sacrificing their own performance to do so. Unfortunately, in bicycle manufacturing – and every other kind of manufacturing, and nearly every other kind of business – this doesn't work. It's the speed of the slowest, or the most difficult part of the process, which determines the overall output. And our bottleneck has always been frames and brazing. That makes the brazers the natural aristocracy of the factory floor.

And this has become part of the corporate culture. If someone wanted to be totally cynical, they could squeeze the brazers quite hard when it comes to pay and conditions. After all, their skills have a very small market outside Brompton. But as well as being immoral, this would send an invidious message to the whole world; if we treated people like that after they invested time and effort in learning a vital skill for our company, how could anyone else trust us to treat them any better? So instead we celebrate their skills. We continue to encourage any full-time employee to follow in Rebecca's footsteps. Having been Abdul's star pupil for years, Rebecca herself is now working alongside him as our training officer, in charge of the academy of brazing that took us decades to set up, but which is going to make sure we can keep on expanding the number of frames we produce in future. If we'd required welding experience or apprenticeships, we might never have found a standard bearer like her for the company, or a trainer who was such a living example to people who might not previously have seen themselves as metal workers.

Both Rebecca and Abdul have their own approach to teaching, but the key characteristic they share is a kind of calmness and kindness. A large part of their training role is being able to talk people down and persuade them to keep going when they're close to tears of frustration. Even though it's a good job and well paid, lots of people who start on our brazing training decide that it's not for them because they don't have the attention span or temper control for it. Conversely, the one thing that the successful trainees seem to have in common is a willingness to listen and pay attention. Rebecca thinks this could mean that women will end up dominating the top ranks of raw lacquer brazers; they are currently 10 per cent of a team of fifty, but we have twenty-seven students currently in the academy, and five of those are women.

This psychological element – the 'inner game' of brazing, so to speak – creates its own kind of bottlenecks. In any activity where you have some employees with extremely high productivity or important skills, then the comfort and mental space of those employees becomes a factor of production; you have to organise things so as to facilitate people being able to sustain their bubble of concentration. For example, every brazer used to have their own station, with their own familiar equipment, in particular their own torches and nozzles. That suited the brazers pretty well. Everyone likes to have their own little home in the workplace, but it's particularly important for brazers, because they need to be confident that everything is going to work and react in exactly the way that they expect it to. Small variations in the thickness of the braze wire or minor impurities can throw them off. Different nozzles for the oxyacetylene torch – even different brands of the same size nozzle – can make the flame behave differently. Not hugely differently, but sufficiently for you to notice and be irritated. The tools might also balance slightly

Rebecca Francis (at left) and Aleksandra Smolna, two of Brompton's best brazers.

differently in your hand, or weigh a little more or less. It's like asking a musician to pick up someone else's guitar. Some guitarists will just pick it up and play without worrying, some will cope but feel awkward, and some will find that everything feels wrong and they can hardly play at all.

So the layout of the brazing department in the old Brompton factory used to be designed around the brazers. Each one would have their own torch set-up and station, and would work on a bicycle frame or part-frame held in front of them on a jig. As the different joints were put together, frames would be passed from station to station and from brazer to brazer.

But this goes against a pretty fundamental principle in designing a manufacturing process, that the stations should be designed around the job. Every time someone needed to change the job they were doing, they had to take the jig off the stand, carry it across to the jig store, store it, find the new jig in the store, carry it back to the station, put it on the stand, then find all the sub-assembly parts, metal pieces and brazing wire, set them out, then sit down and set to work. This would take a minimum of half an hour, assuming everything went smoothly.

And this would happen as often as once a day. Brazers do need to change their tasks frequently, in order to balance the flow of bicycle frames coming through. They also need to take breaks and do something different once in a while, or the repetition itself becomes tedious and does funny things to your head. So you could lose as much as a tenth of a shift's production just from having skilled brazers carrying heavy jigs back and forth and setting up. Added to which, it seemed impossible to design the stations ergonomically for the tasks, or to locate the components next to the places that they were needed.

It took years to change this. Eventually, we found a way to allow each brazer to unhook their brazing torch safely from the 'economiser' (the system that distributes the gas and oxygen) and to carry it from station to station. Today, the brazers move stations when they need to change jobs. Each station has the correct jigs set up for a particular task, with the sub-assembly parts not only to hand at the station but stored in the correct location to be picked up easily. And as the frames are incrementally completed, they are moved from brazer to brazer.

Aleksandra Smolna brazing at her specialised work station.

Doing it this way means we still have an unavoidable bottleneck, but we've managed to improve the flow through it. Possibly more importantly, though, we've now built a system which can expand efficiently. When Rebecca and Abdul's classes graduate to the factory floor, we won't need to add a new station for every new brazer; we will be able to use space efficiently and expand the brazing shop in a balanced fashion.

By starting from the thing which drives the value – completed bicycle frames – and tracing the chain of cause and effect backwards to see where things are being delayed, and then by understanding all of the factors, human and machine, we can get to a place where we know that if we want to double output in X years' time, we need to be starting Y number of recruits on the brazing course today, and planning for when and where we're going to install the next stations and hook up the gas supply. This is what I mean by thinking like an engineer; it's the general approach that matters, not the specific problem. Even when the problem in question is a centuries-old craft that had to be brought back from the brink of extinction, the real solution is not to train people to hold a torch – it's to build an academy.

SPEED

3 INTERLOCKING GEARS

The Brompton bicycle is made of rather more than 1,200 components. Of these, around 70 per cent are manufactured in-house. This is very different from most other bicycle makers, who tend to use interchangeable components made by a small number of global firms.

In late 2000, a few years before I arrived, there was a bit of an adventure at Brompton. Andrew Ritchie found himself in a car park in Nottingham with a van and a cheque book, on a late-night mission to save the company by buying slightly more than a thousand bicycle wheel hubs from the doors of a factory a few hours before it was going to shut down. He hadn't told anyone he was going; there hadn't been time. But without those hubs, Brompton would have had to shut its own doors.

Andrew's dash to save his company set in motion a train of events which kept bringing us back to a single small part on the frame of the bike. We have already seen that one of the good things about manufacturing as a business is that the sequences of cause and effect are laid out in front of you. And that's true, but sometimes the sequences get quite twisted and curve back on themselves, and the knots and tangles can be surprisingly difficult to identify. Nothing has taught me quite as much about problem-solving as a little thing called the 'chain-plate pusher boss'.

The supplier in Nottingham to which Andrew had driven at midnight was a company called Sturmey-Archer. They were a long-established English company, named after the turn-of-the-century inventors of the three-speed bicycle hub gear. This was still their main product nearly a century later, along with an eclectic mix of other bicycle components. (The two kinds of bicycle gears are derailleur gears, which have a choice of sprocket wheels at the two ends of the chain, with a 'derailleur' to push the chain from one sprocket to another; and hub gears, with a set of cogs between the chain sprocket and the rear axle which determine how the force from the pedals is transferred to the hub. In engineering jargon, a 'sprocket' is a cog wheel which engages with a chain, and a 'gear' is a wheel which engages with another wheel – but this last distinction hopefully will not matter too much in the story.)

Hub gears had been going out of fashion since the 1980s, because they tend not to allow such a variety of gear ratios, but they were good for Bromptons because they saved space and they are robust. There were, however, two big problems with them. First, there were only three manufacturers left in the world, two of which made hubs which wouldn't fit the Brompton frame. And second, Sturmey-Archer, which was effectively the only supplier to Brompton, had recently been sold by its parent business to an investment company called Lenark, which specialised in 'distressed' companies and was run by a team of what the City of London used to call 'colourful characters'. The director and main shareholder lived in America, listing his main occupation as 'professional gambler', while the business analyst behind the deal was an undischarged bankrupt. By virtue of their background, the kind of companies which were acquired by Lenark didn't seem to have a very good survival rate.

The death spiral was frighteningly quick. Sturmey-Archer was sold to Lenark in June 2000, and by September there was not even enough cash left for the workers' redundancy pay. The company was later bought by SunRace, a Taiwanese manufacturer, but at the time it very much looked as if the finished stock in the Nottingham factory might be the last batch of three-speed gear hubs in the world that could be used on Brompton bikes. When Andrew heard that there was an insolvency practioner's team on the premises, he drove a van up there straight away. The thousand hubs that he brought back bought enough time for Brompton to negotiate with SRAM, one of the other two manufacturers to produce a usable hub, and to redesign the frame slightly to accommodate it.

While he was picking up the Sturmey-Archer hubs, Andrew met a man called Steve Rickels in the factory car park. The company was by this time down to a skeleton staff, and Steve was really just waiting around for the inevitable end. But he was the head designer at Sturmey, which meant that he was, quite literally, the person in the world who knew most about hub gears. Andrew offered him a job at Brompton there and then. One of the first projects that Steve started working on was a new gearing system.

Steve Rickels, designer of the Brompton Wide Range three-speed hub.

The Brompton Wide Range three-speed hub. When combined with the derailleur, it allows six usable gears with no more weight than a three-speed.

To a large extent, gears were a marketing problem which swam upstream to become a manufacturing problem. People like to have more gears on their bike; if there is a five-speed and a three-speed version, the five-speed will sell much better. Sturmey-Archer made a five-speed hub gear and we used it, but it was annoying. The five-speed hub was achieved by linking two three-speed gear trains together inside a single hub, with the axle from the first hub driving the gears of the second, which then drove the hub that was actually connected to the wheel. It was heavier, for this reason, and it was sluggish and inefficient in certain gears as you were losing a noticeable amount of effort with two sets of gear train losses.

So we had never liked the five-speed hubs, and as it happened we had very low stocks of them when Andrew made his dash to Nottingham. The new suppliers at SRAM did make a five-speed version, but it was really too wide to be usable, and they weren't prepared to make a narrower one just for us. We did not expect SunRace to make it a priority to restart manufacturing of the Sturmey five-speed; in fact it took them some time to start manufacturing hub gears at all, and Andrew ended up sending Steve Rickels over to Taiwan to help them out.

This meant that we would be faced with missed sales as the popular five-speed option became unavailable. It was time to think laterally. Steve suggested that we could combine a compact two-speed derailleur with a three-speed hub gear to give us a six-speed Brompton.

This was a compromise solution which got us back into the market, and it provided customers with the psychologically and commercially important feeling that they had six gears. But we all thought it was possible to do better. The derailleur extended the gearing range of the SRAM three-speed hub at the top and bottom, but in the middle of the range it was really just chopping the same gearing ratios into finer steps.

To solve the problem, Steve ended up designing the Brompton Wide Range three-speed hub. This would be more or less unusable on its own, because the gears are just too far apart, but working together with the derailleur it gave us six nicely spaced steps, covering the same range as an eight-speed hub gear, but with the size and weight of a three-speed. Once SunRace got into its stride, we were able to persuade them to manufacture Steve's design for us, and so in a sense Sturmey-Archer still makes the gears for Brompton bikes. We also went back to the hubs themselves, to see whether they could be optimised. Steve was able to identify that if the hub spent a little more time in the CNC machine having excess metal taken out, it could be made substantially lighter at a competitive cost of about 30p per gram of weight saved. It was also possible to replace some steel parts by more expensive but lighter aluminium parts in the hub, where strength was not a key consideration. The redesigned hub gear came in at 780 g, compared with a Shimano Nexus 8, which, on a like-for-like basis, weighs 1.7 kg.

This is one of the big advantages of getting things custom designed. Components made by major manufacturers are designed and made to be fitted to the average bicycle, with goals of achieving safety and long life at a reasonable cost. This doesn't mean that they're 'worse' than our components – that isn't really a meaningful statement to make. It's just that they aren't optimised for what we want to do with them. There is no reason to spend a small amount of money to save a kilogram of weight on most bikes, because people don't often pick them up and carry them. When you're pedalling along the road, most of the weight you have to shift is your own body; a single kilogram doesn't make much difference to that. So a process which removes weight but costs money is generally not worth bothering with for most manufacturers. But for our purposes, in terms of pence per gram, once the investment in design and tooling was amortised, this was an amazing trade-off.

Solving one problem, however, has the tendency to create another. Let's take a close look at that rear derailleur, because Steve Rickels's masterpiece included a few little components which were to cause headaches for the next decade. As the Sturmey-Archer saga demonstrated, the thing about components is that they fit together. If you run out of hubs for your wheels, it doesn't matter how many frames you have in stock; production stops. We try to keep a reasonable inventory of components so that the assembly line won't come to a halt – this has been a policy dating back to well before the Sturmey episode, and it served us well when our Brexit contingency stockpile suddenly had to do duty as a coronavirus contingency stockpile. But not only do components all have to come together, they also need to fit together just right. Consider, for example, the derailleur gear.

A derailleur gear works, as briefly mentioned above, by moving the chain from one sprocket cog to another cog of a different size. This has to happen smoothly, while the chain is moving, in response to the pull on the gear cable from the lever on the handlebars. The way this is achieved is to have the chain go round a pair of 'jockey wheels' (two more small sprockets, mounted on the bottom half of the chain which isn't under tension while you're pedalling). The jockey wheels are arranged so that they can be pushed from one side of their axle to the other, carrying the chain with them and changing gear.

The bit that pushes the jockey wheels is called, prosaically enough, the 'chain pusher'. When it's bought as a spare part, it looks like an assembly of two components. One of them is a piece of injection-moulded nylon plastic which holds a bearing, connects to the cable and moves from side to side when the gear lever is moved. The other is a stainless steel plate which attaches to the plastic component, with two wings sticking out. This is the bit which actually pushes the jockey wheels. If you look closely at the plate, you can see a little bit of discoloration of the steel, which shows you how it was produced; it's stamped and formed under extremely high pressure. (You can, if saving ten grams of weight is important to you, buy a pusher plate machined out of titanium by one of the enthusiastic aftermarket manufacturers in Asia.)

When it's considered as part of the bike, though, the chain pusher has three essential components, not two. The injection-moulded part has to be fixed onto the frame, and the fixing point has to be absolutely solid. That is achieved by screwing it into a small hexagonal bump (a 'boss') which has previously been brazed onto the frame.

The plate has to touch the jockey wheels in exactly the right place in order to operate smoothly. The pusher mechanism only has a certain amount of travel built into it, and this will only shift the chain across if the contact between the 'wings' of the pusher plate and the edge of the jockey wheel sprocket is happening in exactly the place it was designed to happen. The difference between right and wrong isn't drastic – it's about the order of a single millimetre. But if the chain pusher isn't aligned correctly, the gear change won't be smooth – the chain will go click-click-click-click as it tries to get from one sprocket to the other. The rider will quickly notice; it is a very annoying noise, and in the extreme the gears won't change. It will be obvious that the bike is faulty.

The infuriating thing about this kind of problem is that it can be very difficult to pick up in quality control. To begin with, you have

Sturmey-Archer gears are now produced by SunRace out of their factory in Taiwan.

the usual issue of stacking tolerances; the boss, pusher and plate are fitted on top of one another, so the total possible variation, in the worst case, is the sum of the three. The main problem with the boss tended to be that there might be a bit of excess metal around the joint, affecting the fit of the pusher assembly. The pusher itself had a certain amount of tolerance in its shape and in the fit of the bearing, and this would interact with tolerance in the stamping of the plate, which was a component bought in from an external supplier; both parts could also have been fitted just a little bit off-centre. If it was always the same flaw in the same component, it could have been much simpler; just train the brazers and inspectors to be careful. But with three parts involved, it was that much more difficult to track down exactly what was at fault.

And these weren't the only tricks that the chain pusher had to play on me. It is, of course, a moving part. That means it is subject to wear. It's inevitable for the wing-shaped plate in particular to get worn, as its job is to push against a moving object. And the moulded plastic part has a bearing in the middle and is subject to stresses; over time it's reasonable to expect that the assembly will get a little bit out of true. For this reason, it has some adjustment screws which can be loosened or tightened with an Allen key.

When the chain starts going click-click-click, the cyclist's first port of call might be a bike repair shop. If it is simply a matter of adjusting the screws, this is normal maintenance. If the screws can't be adjusted sufficiently and the chain pusher is worn out, that's also fair enough; the bike shop will sell them a replacement. One of the good things from a business point of view about making so many of our own components is that we keep a lot more control over the spares (and we'll look at this again later in the book, as it affects how we deal with distributors and retailers). But occasionally in the past, people found that the adjustment screws had reached the end of their range even though the part was clearly not worn out.

What seemed to happen was an even more subtle version of the problem of stacked tolerances. Any individual part might pass the

The SunRace factory in 2004.

checks, and the whole assembly might also pass – but at the edge of its tolerance. This could mean that after only comparatively little wear, it was then out of true and the customer experienced a fault. Whether this would show up or not was largely a matter of luck; if the cyclist had a soft hand on the gear lever, or didn't pedal so hard, or if they just happened to be lucky, they might not get so much wear and never know about the problem. An intermittent fault – one that doesn't always show up in the same way under the same conditions – is always more difficult to diagnose and treat than a consistent one. In this case, all we could really do was to reduce the permitted tolerances on the assembly as a whole, which meant adjusting the jigs, no-gos and pokey-yokeys and making sure that the staff were trained and alert to the possibility of this problem.

Time went on, and a new problem showed up. We used jigs and gauges to ensure consistency and maintain tolerance. But after producing a few thousand parts, the jigs themselves began to wear. If you think back to my DIY tip for home-made jigs to accomplish domestic tasks, it's obvious that it wouldn't have been a permanent solution. The cotton-reel drill jig would be all but destroyed after a few holes. But the same thing is true even of a precisely engineered, carefully manufactured production jig. It won't fall apart, but over time pieces will wear by just a little bit. Things that are meant to be rigid will bend or loosen slightly. And it wasn't just our own tools that we had to worry about. For example, the stamping tool which bent the stainless steel to make the pusher plate also wore down a little over time, which we had to identify by conducting quality checks on suppliers' parts as they came in.

So as we ramped up production, there were problems of scale. But in many ways, by the time – roughly ten years later, in 2017 – that I was able to finally tell myself that the chain pusher no longer needed permanent attention, it was the scale we had grown to that was actually providing the solutions. In other words, it wasn't because of any one single thing that we did: instead, the whole system had changed in such a way that we no longer created problems for ourselves. As Brompton reached the point of making and selling tens of thousands of bikes a year, we could start investing in things like replacing the jigs ahead of schedule, or maintaining the tools proactively before they wore out. And this leads me to the real lesson here, because this chapter hasn't just been a shaggy-dog story about how we spent ten years trying to make marginal improvements on the assembly of a component that costs less than £20 as a spare at retail.

The point is that fixing problems doesn't work. If you go into management thinking that your job is to solve problems, you will retire unhappy, because every day brings new ones. Metaphorically, every jig you make will eventually bend, every gauge will drift out of calibration and all the problems you thought you had solved will recombine to create intermittent faults. The ability of any complicated system to generate problems is greater than your ability to solve them. The only way to avoid facing this fact is to strangle your company, preventing it from getting any bigger than your personal capacity to handle problems. You can even fool yourself for a while that this isn't what you're doing, if you hire people like yourself to walk around solving problems on your behalf. There might be a one-in-a-hundred chance that you won't go mad with the stress of delegation. But eventually, if you try to do things this way, the whole structure will fall apart under its own weight, because trying to solve problems one by one isn't a viable system. You can't win a game of whack-a-mole by buying a bigger hammer.

Instead, what you have to do is build processes and empower people to find and address the problems themselves. That's not quite the same thing as having an information system. Imagine if the human body's immune response worked by sending you a text message every time you inhaled a bacterium and telling you to go to the pharmacist to buy the right medicine. You would be overwhelmed with messages and most likely dead within a year. The way in which viable systems work, when they are set up well, is for the solution to happen almost as a direct, causal consequence of the problem. So a proper proactive maintenance system means that when a jig begins to go out of true, it starts a chain of actions which ends with that jig being repaired or replaced. A rejected consignment of parts from a supplier starts a sequence of inquiries that ends with changes being made to ensure that the next consignment will be back within specification. Every time feedback from customer returns identifies an issue, we need to make sure that we learn from it and solve the general case, not just do the specific thing to make that symptom go away for that customer.

And the really crucial point is that the fixing processes need to happen at the level of the organisation that is closest to where the problems themselves arise. You need to trust and empower the people who are doing the job to identify things as they go wrong. If I as a manager find a problem, that ought to start me worrying; why wasn't this caught before it reached me? Throughout this book,

I'll argue that at Brompton's current size, its real competence isn't building bikes. It isn't even building machines to build bikes, as I used to say in interviews in the 2010s. Now, it's building systems which allow people to build machines and build bikes.

The job of managing a manufacturing business isn't to solve problems: it's to develop processes and people to solve problems.

By now, we've come quite a way in terms of describing how training as an engineer and working with some great people have taught me how to make bicycles, and how I've tried to generalise those ideas and lessons into an overall method. We've started with the concept of a jig and a *poka yoke*, and moved on to thinking about building systems and making problems solve themselves. We've also looked at bottlenecks and begun to understand that things are designed for a purpose, and that the way you design components and processes affects the kind of control you have over them. Each one of these insights was pretty hard-won for me. But once the problems had been solved, usually by trial and error, the lessons were learned. And after a while, I was able to apply the same ideas to some really complicated issues, and to things which weren't anything like as easy to keep under control.

4 QUALITY AND CONTROL

> *Brompton has dozens of outside suppliers, mainly for commodity products like steel tubes and bolts, but also for some more sophisticated manufactured components which are not practical to make in-house.*

Infuriating though it was, trying to perfect the chain pusher never put me in any physical danger, which is more than I can say for some components made by outside suppliers. You have a lot less control of what's happening outside your own factory, and trying to manage that problem has seen me freezing by a roadside in Poland and trying not to get molten metal spilled on me in Walsall. One of the big benefits of doing so much of our own manufacturing is that it allows us to build really strong relationships with the relatively small number of suppliers we do have. So here are three stories of dealing with other manufacturers. One we outgrew, one was a disaster avoided, and one was the best of experiences.

Often a major advantage in doing things yourself is getting a solution that's exactly right for you, although that isn't always the case – after all, the Brompton Wide Ratio gear hub is made in Taiwan. But that's an exception that proves the rule. We want to optimise the design of each component so that it contributes to a product that's ideal for the task it was designed for. An indirect consequence of this is that we create a lot of things with unique qualities, which the competition can't copy.

A component which has 'Brompton-ness' may be simple or complicated in its design, but we'll have learned about it in the process of making it. Our staff will have been trained in the ins and outs of making the thing, and our jigs will be on their tenth generation. The thing itself might have been redesigned a few times. If you go through the process of detailing all your requirements to a supplier, then monitoring their output and providing feedback into their process, over time the know-how relating to that part gradually relocates from your factory to theirs. Their staff are trained in it, and after a while their jigs are the ones on the twentieth iteration. That's the sort of knowledge you don't want slipping out of your hands. Not only would the part no longer be unique to us, we'd also lose the opportunity to make future redesigns.

On the other hand, for some things there's no real Brompton-ness in the know-how because there's nothing we can learn. For example, there are several formed pieces of wire on the frame of a Brompton which we call 'cable gatherers' because they keep the brake and gear cables out of the way of the folding action. There's nothing unique to the clip's design; it's just made out of a piece of wire bent into shape. When we produced these in-house, it was a matter of getting someone with strong forearms to cut five hundred pieces of wire, putting the first bend in them, resetting the press, making another five hundred bends, and so on until a stock of components was produced. It took a day or two, and some of the bends wouldn't come out quite right, meaning that there were always some rejects. There's really no other way of doing it at the scale on which Brompton operates.

But CNC wire-benders exist; machines that can produce a thousand cable gatherers in half an hour, each one of them perfect. We couldn't justify investing in a machine like that, but a specialist wire-bending firm can. Similarly, the steel tubes which come into the factory are cut off at a right angle at the ends, but the brazers need tubes with ends that will fit flush with another tube. There's no special design here; it's just a curve following a circle of a particular diameter. There's no advantage for us in cutting these 'cods-mouths' ourselves by hand when a specialist steel-cutting firm has invested in a laser cutting tool that can produce them quickly and perfectly.

A worker 'fills' the hub with spokes before the wheel is 'laced' and 'trued'.

Here's another clue from the language of industry and engineering – the words 'quality' and 'control' come as a pair. The more you need of one, the more you'd better have of the other. There are some quite important engineering principles here which are best illustrated with a couple of stories.

The socket-brazed hinge that I described in Chapter 1, for which we had to buy new CNC machines, wasn't just a new brazing and machining process. The actual component itself was made differently – rather than using forged mild steel, it was made out of whiteheart malleable cast iron. Without straying too far into metallurgy, this is iron which has a 'white heart' because the surface has had all the carbon oxidised out of it while the centre is composed of 'pearlite', an extremely strong crystalline structure of iron and carbon. It's 'malleable' because it's been heat-treated to reduce the brittleness of the metal crystals, and you make things out of it by 'casting', the process of pressing a mould into a pile of sand and pouring molten metal into the cavity created.

We mentioned before that the basic method of sand moulding is essentially unchanged since Roman times, but some foundries have moved on more than others. The foundry which we used for the whiteheart malleable iron socket hinges in 2003, when they first came in, was Deeley's Castings in Walsall, on the outskirts of Birmingham, and it seemed as though the only thing to have changed there since the days of the Peaky Blinders was the menswear fashions. It was an extraordinary place to visit and watch. At one end of the shed was an electric induction furnace, no bigger than the kind of wine barrel that people sometimes make outside tables from, in which scrap metal was melted. One of the foundrymen stood over the cauldron, pouring carbon pellets in until the composition of the molten metal was just right. The windows were dark with accumulated smoke and the floor was caked in decades' worth of dirt and spilled casting sand. Nobody wore a visor or breathing equipment, hardly anyone even had hearing protection. A frame the size of a small coffee table stood on the floor, into which sand was poured, and then the templates for the two halves of the hinge were pressed in.

A foundry worker then packed more sand around the pattern by hand before closing the two halves of the mould together. Next, two workers poured molten metal out of the cauldron into what was effectively a bucket on a pole, looking like a long ladle, and carried it over to pour into the mould. Five minutes later, the iron had

Molten iron is poured into the ladle at former hinge supplier Deeley's Castings.

solidified and it was time to scoop the sand away. The foundrymen first hit the metal pieces with hammers to break off the 'runners and risers' (the excess which had collected in the channels left in the sand to allow metal into the mould and air out), then picked them up with tongs and put them onto a trolley, ready for the heat treatment to make them malleable rather than brittle.

It was fascinating but also frightening to watch, particularly for a young manager whose only experience had been on the relatively civilised factory floor at Brompton and before that the scientifically controlled environment of a chemicals plant. In many ways those foundry workers deserved a lot of respect – you can only produce quality output from a process like that by having an extremely high level of craftsmanship. The skill of the ironworker pouring carbon into his cauldron was close to that of a Michelin chef, and everyone involved was a master of their own trade. But this meant that it would be hard for them to grow with us and maintain quality, because everything was dependent on this small number of artisans.

We tried our best to work with Deeley's. We asked for more paperwork and certification of the materials before the pour, we asked for samples, and so on. But it's not really possible to bolt modern quality control onto a fundamentally pre-industrial process. However good the craftsmen were, we were always worried – what if the guy who knows how to do it is ill? What if someone gets married and the whole factory goes out to celebrate? In the end, we changed our supplier to Swan Shephard, a larger company in the area which specialised in big one-off pieces like sculptures and which had moved its mass-production operations to Poland. Deeley's Castings unfortunately isn't in business any more; its former site became a trading estate and is now owned by a car auction company. Its name survives only on the front of replica vintage football shirts which are still on sale commemorating the 1983–84 season when Deeley's sponsored Walsall Football Club.

Visiting the Swan Shephard plant in Katowice was a revelation. I had thought that it might be cold, but I was ready for something like a British winter, rather than –20°C. We even ran out of petrol on the journey from the airport to the foundry, to really bring it home to me how extreme the conditions were. Everything was built on the former communist scale. The Deeley's workshop in Walsall had made about eight hinges at a time, and struggled to keep up with our orders. The Katowice plant put moulds together on a massive machine (called a 'DISAmatic', after the Danske Industri Syndicat who patented the process) in two blocks of thirty-six hinges each; it took them two hours to produce a month's worth of hinges. We were barely ordering enough parts to make it worth their while to start up the machine. Their furnace was the size of swimming pool, with a magnetic crane to drop in whole cars and ovens, and continuous electronic measurement to ensure the quality of the iron. I stood watching it, shivering despite the heat of the furnace (and despite the coat I'd borrowed from Przemek, the Swan Shephard metallurgist).

Working with an operation like this allowed us to raise our own game and to start making changes to push the DISAmatic to the limits of what it could achieve. We discovered that it made sense to create the templates for casting the hinges slightly over-size, and then machine away the excess metal. This allowed a better flow of metal into the casts, which meant in turn that the carbon would oxidise away more readily and the material quality would be higher. We redesigned the runners and risers for similar reasons, and now the hinges come out stamped with a batch number so that we can track them for our own quality control. I still go out to see Przemek from time to time, although preferably in the summer.

This need for control over the quality of materials has to extend to everything; you can't make good bikes out of bad stuff. Every tube in a Brompton bike is made from a slightly different steel alloy, with slightly different properties. But it's not only a matter of different alloys; which supplier you buy your steel from matters too. Anyone who bakes bread will be aware that different brands of flour perform differently, and that designations of 'hard bread flour' or 'type oo' and the like are quite broad categories which admit significant variation.

Making steel is the same, and the tolerances can be wide. Suppliers who take care to produce a consistent product with as little variation as possible from the published specifications can charge a premium. We pay that premium, because in our product, increased variation has a direct impact; it means you have to design for the worst case in which everything is at the wrong end of its tolerance. That means extra weight, because the margin of safety you bring in for that worst case is made up of extra metal.

We used to check up on our steel tube suppliers with a wonderful device that Andrew invented; another real piece of DIY genius. Every batch coming in needed to be tested, as one load might contain tubes for thousands of bikes – enough to seriously damage the company if they needed to be recalled. But a 'yield strength test', the standard means of testing the strength of steel, would not really have been possible to fit into the process. The test consists of cutting a standardised shape, then using a machine tool to literally pull it apart, measuring the force applied to it and plotting a curve as it deforms. The press you need for this is big, slow and expensive. The suppliers are meant to regularly test their output against specification, and we would always ask for the certification of this testing to be sent along with every batch. Every now and then we would send samples off to a specialist company to make sure, but what we really needed was a way to systematically audit the steel as it came in.

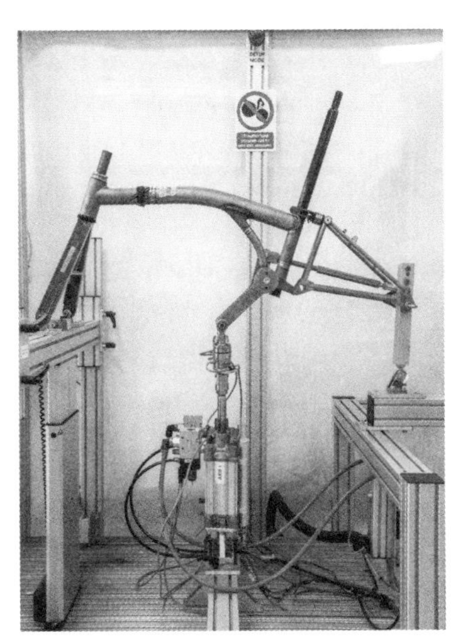

A frame testing machine can simulate the stress of years of everyday riding.

Andrew's solution was the 'squish test'. He reasoned that you could cut the last two centimetres off the end of a tube to form a small ring, measuring it accurately so that you knew exactly how much steel you were testing. This could then be squashed on a considerably smaller press than for the yield strength test, measuring the force which was needed to bend it, then bend it a little more, and finally to take it to the point of permanent deformation. Plotting a curve of the force applied and the degree of squish in the ring would give a curve corresponding to the actual yield strength curve as measured in the standard way. Every six months we would continue to send off a sample to check the calibration of the squish-testing kit, but in a way, the precise numbers didn't matter – we could compare the curves from new batches against our previous results and see if they were in line or way off.

On at least one occasion this saved us. Alan, the technician responsible for testing a batch of main frame tubes, found that one batch which had come in was 'like butter'. After repeating the test three or four times to make sure we weren't mistaken, we had to call up the supplier. This wasn't the easiest conversation to have; they were a big steel company, we were a small customer and the certification seemed to be wholly in order. But it turned out that there had been a shipping mistake, and they had sent us a batch that was meant to go somewhere else.

If it hadn't been for the squish tester, we could easily have made 2,000 bikes which would have fallen apart three years later – so that was a disaster avoided. Interestingly, it was the supplier's bad luck that the delivery mistake went to Brompton. If it had gone to almost any other bike manufacturer, probably no one would have noticed. Although the tubes were way out of tolerance for our purposes, they were not so far out as to have exceeded typical safety margins. If they had gone into a kid's bike, for example, the frame would have been used occasionally and recreationally for a couple of years, then grown out of and scrapped or recycled without anyone ever knowing it was faulty.

This is the sort of fundamental engineering principle that shapes businesses. Not only does form follow function, but the processes that you design are themselves determined by both the function and the form. Everything starts with the fact that the bicycle is how it is, and so it has particular requirements in terms of the variation we can tolerate. Tolerance is a measure of quality, and quality has an intimate relationship with control, because the extent to which things are out of your control defines the worst case that you have to provide for. These are the sorts of principles that I learned from Andrew Ritchie.

Our happier experience, and the way that ideally I'd like all our supplier relationships to go, was with one of our best suppliers, the Hesson family, who ran Knight Cycles of Dudley. This relationship was a bit different, because they supplied our wheels up until 2015; the wheels intrinsically have a lot of 'Brompton-ness' to them and a lot of company DNA, but we didn't have the expertise, or the physical space, to make them in-house. Mike Hesson was a cycle enthusiast of the old school, a wonderful, charming man who used to have his own custom cycle-building business. By the time I arrived at Brompton he had stopped making frames, but his custom wheels were still in demand. Making a bicycle wheel is a skilled job; the spokes have to be 'laced' into the rims, and 'trued' to adjust the tension on each spoke so that the wheel is perfectly round and spins smoothly. Mike and his daughter Angie would do this job by hand at their workshop, with Mike riding twelve miles from his home in Wolverhampton each day, whatever the weather.

Even at the production levels when I joined, the demand for Brompton wheels was stretching their capacity. After all, 5,000 bikes a year is 10,000 wheels, plus maybe another thousand for spares – about forty wheels a day. Realistically, there was no way they would be able to keep up if my plans for increasing production were to go anywhere. Wheel production was going to have to be mechanised.

Mike was sceptical about this; he didn't believe that any machine could do the job as well as he could. His view needed to be taken seriously, because he was a good person who knew a lot and the quality of the wheels is not something that we can afford to compromise on, while I didn't really know much about wheel lacing and nor did anyone else at Brompton – that was why we were buying from an outside supplier in the first place. The machines that can automatically lace and true bicycle wheels were made by Holland Mechanics in Amsterdam, so we took Mike on a trip to the Netherlands to look at some of them.

The automatic wheel-truing machine, a masterpiece of automation.

Holland Mechanics is itself a family business, run by Wouter van Doornik. He and Mike hit it off pretty quickly; they both care about quality more than money, and Mike was gradually brought round by the complexity and sophistication of robotic wheel lacing, and particularly by the laser sensors that allowed the machine to spin a wheel and deduce what adjustments it needed to make in order to true-up the spokes. We came up with a deal whereby Brompton bought one of Wouter's machines and leased it to Knight Cycles, with our lease payments coming in the form of cheaper wheels. Mike ended up being able to grow his total revenue, while providing a lot more wheels and helping to do our R&D.

Keeping the Hessons in the loop very soon proved to have been the right decision. Of course, all of the intricate demonstrations we had seen in Amsterdam were done on full-size wheels. Nobody had used a Holland machine for 16-inch wheels before, because that size of wheel was generally associated with children's bikes, and they aren't considered to be worth trueing properly. As we were all taught at school, the area of a circle is given by πr^2, which means that the amount of room the machine has to work in reduces as the square of the difference in the radius. Added to which, shorter spokes don't bend as easily, so it's harder to adjust them. And the Holland Mechanics Intelligent Lacer is a delicate piece of engineering itself, and it doesn't take kindly to being bodged and pushed about.

Mike's knowledge and expertise meant he could settle the lacer down and give feedback to the Holland factory, so Wouter and his team were able to adjust and evolve the machine for us in order to handle the smaller wheel size. Knight Cycles were still making our wheels by the time we moved into the new factory in Greenford in 2015. When we finally had the space and capability to bring production in-house, it was largely because Mike wanted to retire: he had reached the age of seventy-five and didn't want to buy a second Holland Mechanics machine and employ more staff to keep up with our demand. Even so, we needed to agree a two-year handover process after we took the wheel-building in-house so that he could pass on his expertise in handling the robots.

In order to get that kind of relationship with a supplier, however, you need to put the hours in. This was one of the ways in which I first tried to make myself useful when I was hired. Andrew wasn't very keen on travelling and tended to presume that other people either had the same standards as him or were no good and could be dropped. I made a point of getting on the road, going to Taiwan or Walsall or the paint shop in Wales, and it paid off. Quality and control aren't just about measurements – a lot of the time they're about how far you can trust the people involved. That was a lesson that became more and more important as the company grew, and we'll come back to it later.

5 DEVELOPING THE SUPERLIGHT

> *The titanium alloy used in Brompton bicycle frames is called 'grade 9'. It's also known as Ti-3-2.5 because it contains 3 per cent aluminium and 2.5 per cent vanadium. This isn't quite the strongest alloy possible, but nearly. (Grade 8, or Ti-6-4, is used in aero engines, but is so strong that it's difficult to draw it out into tubes.)*

I always used to complain about the 'Aladdin's Cave' jumbled approach to organisation and component storage in the Brentford factory, but it was also an Aladdin's Cave full of hidden treasure. While I was looking around one day in 2004 and trying to tidy things up, I discovered a titanium rear frame for a Brompton. It was not particularly inspiring; the measurements were out by some way, and the chain pusher plate boss had been attached at completely the wrong angle. But it was still a titanium bike frame, and given how obsessed with weight we were at the time, I couldn't help finding it interesting. Andrew had clearly put considerable work into it at one point, and I was surprised he'd never talked about it. I took it on as a private project, and after a lot of filing, bending and bodging, I had something which I could put wheels into and ride. I still have it – the first ever titanium Brompton.

This is the last chapter to deal with manufacturing, and in it I will try to bring together several related ideas and lessons, and show how we applied them in challenging and complicated contexts.

Mainly, it's about the processes we went through in order to start making titanium frames. And this time we won't make the mistake of solving problems in the wrong order. In order to start thinking about the materials and trade-offs, we are going to start with the output and the value created. So we start with titanium, because the purpose of this exercise is to reduce weight.

Ed Donald, our first marketing manager, always tried to remind us that 50 per cent of the potential market were women, and that female customers would care more about the bike's weight than almost anything else we might do with it. This wasn't as obvious as it might seem; in fact, weight isn't particularly important to a lot of bike manufacturers. When you're riding a bike recreationally, like a BMX or a road racer, what matters is the weight of the whole

The Superlight S2LX, one of the early part-titanium Brompton models.

system, bike plus rider. A kilogram's difference is barely 1 per cent of this total weight; although cycle enthusiasts always like to reduce weight, that sort of reduction is not very noticeable in everyday use. So most bicycles aren't built around a strength/weight trade-off, because there would be little point in doing so; you'd be adding expense for something the customer might hardly notice. Folding bicycles, on the other hand, are meant to be carried.

At between 10 and 15 per cent of your body weight, there's a range of things that the average person perceives as 'heavy to carry'. Around the top end of this range you have something like a mail deliverer's fully loaded bag. According to the postal unions, the bag shouldn't weigh more than 16 kg at the start of a round if it's being carried ergonomically by a fit person who's used to the task; less if the round involves hills or stairs. At the bottom end of the range, the average ten-month-old baby weighs a little less than 9 kg. That's closer to the sort of weight that a physically healthy parent would, without necessarily looking forward to the experience, be prepared to carry for extended periods of time, like a shopping trip or a visit to the zoo. Anything much lighter than that moves into the realm of a shopping bag or a student's backpack; something you might carry all day without thinking too much about it. Between the upper limit of what's possible for a fit 90 kg man to carry, and the upper limit of something which might be comfortable for a 60 kg woman, there is a range of about seven kilograms.

Before the Brompton, the few folding bikes that existed on the market were very much towards the loaded mailbag end of the scale. They were mainly designed to be put in the boot of a car and transported to the start of a recreational ride. It was just about possible to carry them, but not comfortably; in the language of the earliest laptop computers, they were 'luggable' rather than 'portable'. For a folding bike to be a realistic everyday tool for a commuter, it needed not only to fold up small enough to take on public transport but also to be ergonomic when picked up – meaning that it needed to weigh not too much more than 10 kg. The only bike which came in around this weight was the Bickerton, but that had problems with rigidity – indeed, it was Andrew's frustration with the 'blancmange-like' ride quality of the Bickerton which initially inspired him to invent the Brompton.

The key problems to solve in a folding bike are rigidity and strength, and the easiest way to make something strong is to add more metal to the structure. Conversely, the easiest way to make a bike that weighs less is to remove metal from it, and this will, other things being equal, make the structure weaker. Nearly all the skill and intellectual property in the Brompton (apart from a few purely aesthetic flourishes) is in the design and manufacturing techniques which allow us to push this trade-off to somewhere near the optimum, designing strength into the parts which need it and cutting metal out where it's not adding anything.

One of our ideas here was the creation of the two-speed Brompton, a real triumph of lateral thinking and realising that less can be more. If you take the six-speed edition and remove the Wide Range hub, then you have a bike with two derailleur gears – one for going up hills or getting away from the lights, and one for cruising along. It's not ideal for every city, but for people whose commute doesn't include Mont Ventoux, it saves almost an entire kilo of weight, and the bike costs less.

But in order to really get weight down to the comfortable-to-carry range for the standard models, you need to start addressing the big pieces of the frame, and this unavoidably makes things more expensive. We try to think in terms of a trade-off (pence on the price per gram off the weight), but it's not incremental; big changes have big cost implications. And a lot of potential solutions can quickly be written off as unworkable provided you keep thinking backwards from the way the product is used rather than being entranced by cool technologies and trying to find uses for them.

Carbon fibre, for example, is a great material for some kinds of bicycle frames – it's incredibly light and it can be moulded into shapes that would be difficult to achieve with metal tubes. But for a

Andrew Ritchie's 'fatigue rig' allowed design changes to be tested for their effect on frame strength. Nowadays this task is mainly handled with computer modelling.

workhorse bike, it has some less-than-ideal properties. A knock or scratch can produce small nicks ('stress raisers') that you might not even see, but which could cause the frame to fail explosively with no warning. Racing bike enthusiasts who really value the weight trade-offs may be prepared to carefully inspect their frames every weekend, but it's not practical to do this with a commuter cycle that's meant to be ridden in cities and put into luggage space. We are starting to use carbon fibre in components where we think the risk can be kept under control – forks and handlebars on the very lightest models – but a full carbon frame doesn't suit our users.

Titanium, on the other hand, is interesting. It's light and it's got just the right amount of stiffness and flexibility. It is resistant to corrosion because it forms a thin but strongly bonded layer of chemically stable oxide on any surface that's exposed to air – in fact, this layer is so stable and so quick to form that you need to swamp newly cleaned titanium surfaces with argon gas in order to be able to weld them. It is more expensive than steel, but not so wildly more expensive that you couldn't conceivably make a consumer product out of it.

The prototype titanium rear frame that was lying around in the factory stores was not the result of such a logical process of consideration, though. Andrew had been contacted through friends of friends by Rupert Wilbraham, one of those go-getting young men who went off to the former Soviet Union in the immediate aftermath of its collapse in 1989 to see what adventures were available. Rupert had been running a project to locate and catalogue wrecked aircraft from the Second World War and picking up interesting contacts along the way, one of whom led him to a company called Rapid. Rapid was based in Korolev, a former Soviet 'science city' on the outskirts of Moscow which used to be the centre of the Russian space programme and before that a major centre for missile and tank production. As with so many companies in that time and place, it was made up of a lot of talented scientists and engineers who were looking for applications for their skills in the post-Soviet world. Having worked with titanium in the aerospace industry, they were seemingly brainstorming what else it could be used for – nowadays they also make specialist equipment for circuses.

Wheel hub forging from solid aluminium at Chosen, Taiwan.

Shortly before I arrived in 2001, Andrew was sufficiently interested to give Rupert a rear frame to take to Rapid and see what they could come up with. But things seemed quickly to have gone wrong. The Rapid team tried to copy a Brompton frame, but they were working on

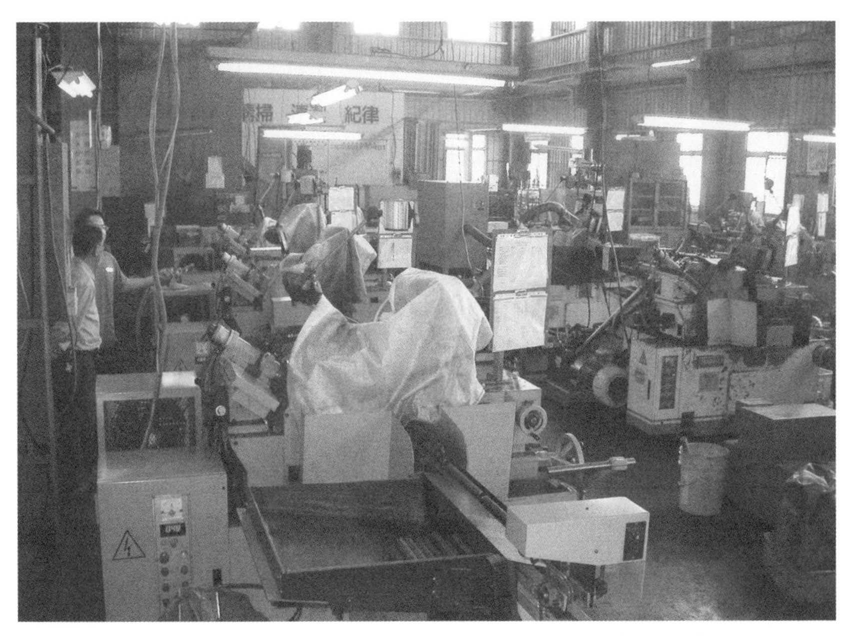

Ain Tec's factory in Taiwan, where titanium forks for some early Superlights were made.

spec to make a one-off without jigs and tooling. It's possible to make something that looks the part that way, but without jigs it will never be precise or technically viable. Rupert enthusiastically presented what he regarded as a proof of concept, but met the same fate as so many others who had made the mistake of bringing Andrew a half-baked piece of work. The copy was inaccurate in a number of ways, and Andrew, not entirely unreasonably, took this as a sign that the Russian factory wasn't able to deliver the standards needed and that the project wasn't worth pursuing.

And there it might have rested. But by pure coincidence I was friends with Rupert's half-brother. (It's a small world; Rupert's other brother went on to marry my sister.) I met Rupert socially shortly after I arrived at Brompton, and he told me about his past involvement. That was why I had set about finding the lost prototype, and getting it working persuaded me that it was worth giving it another try. After a fair amount of persuasion, Andrew agreed that the project was worth revisiting, and quite soon I was making my own trips out to Korolev. Having discussed all the technical aspects with the operating and engineering staff, I had to make a deal with the owner of the factory, and this involved a lot of hospitality. For the Russian owners, the deal wasn't just about money: they wanted to be sure of my character, and their approach to due diligence involved endless rounds of vodka, accompanied by huge orange fish-eggs served on local bread. By the time they were convinced that they could do business with Brompton, I was very badly feeling the effects; it was a tough journey back to Moscow.

In a matter of months, we realised that it could in fact be done; if we could ensure stable supplies, we could get a production Brompton into the shops, with a titanium rear frame and forks and a two-speed derailleur, and weighing less than ten kilograms. That would be a bike that would really deserve the name 'Superlight'. We just needed to do a lot more work and then take a risk. The work part took a year. First we had to transfer our blueprints, including a lot of Andrew's original pencil drawings, into a CAD system compatible with the Russian one. Next we made a number of special jigs for the titanium frame, which Rupert would pick up and take away in large suitcases every time he visited the UK. It took a number of iterations, and I got more used to the vodka hospitality, but eventually we were ready to face up to the risk.

The supplier issues in titanium have historically been quite basic; the problem is not quality control, it's persuading someone to deliver to you reliably. Many of the big producers of titanium 'sponge' (the basic mass of crystals purified out of the ore, from which you can start to make ingots and tubes) are in quite remote parts of China, Ukraine and Kazakhstan. It's a specialised industry and you used to have to deal with some fairly extraordinary people and places. You also needed to adjust your level of aversion to financial risk.

There's a curious, light-headed feeling that takes over when you decide to commit a very large sum of money to a project from which you're not sure you will get anything back. Andrew and I argued a lot about spending money, but the decision we took together in 2005 was probably the biggest and craziest risk of the lot – a bank transfer to Rapid over in Korolev. We had helped them out with some jigs and tooling to overcome the quality control issues which had affected the first prototypes, but it became clear that in order to get our first proper consignment of titanium frames made, we would need to go into the money-lending business. Rapid were working in the post-Soviet economy, not all that long after the political chaos that first brought Vladimir Putin into power. The Russian banking system was not in great shape and everyone wanted to be paid up front. We could have placed a £100,000 order with them for frames (a huge amount for us at the time; the company's annual profit that year was only around £300,000), but that wouldn't give them the cash to buy the titanium tubes. The only solution was for us to finance them.

This was not at all a comfortable thing to do. The business press was full of stories of people getting ripped off in the chaotic legal system of Russia. I had met the owner of Rapid, drunk vodka with him and all that, and I liked him, but how much did we really know about him and the company? And how much did we know about the people who were meant to be supplying the titanium? Andrew and I discussed this for some while, before basically talking each other into a rush of adrenaline and sending the bank transfer. At that precise moment, we couldn't be sure whether the material for our frames had even been dug out of the ground. But Rupert and his team were as good as their word – our trusting nature was rewarded and we were able to start listing the Superlight models in 2006.

Our other titanium supplier was Ain Tec, in Taiwan. After our previous bad experiences with Neobike, we were initially reluctant to send too many jigs out to Asia, but the front forks of the bike had a lot less Brompton-specific engineering in them than the rear frame, and Ain Tec were a good and reliable company. This diversified us to a certain extent; for a variety of reasons, the supply of frames from Korolev was subject to occasional delays and interruptions, so the fact that we didn't have to worry so much about the supply of titanium forks meant we could keep Superlight production going without having to hold very large precautionary stocks. Ain Tec subcontracted the manufacturing to their specialist subsidiary in Baoji, in mainland China. Naturally, I felt I had to visit.

If the Rapid factory in Korolev had been a slight culture shock, the titanium plant in Baoji was something else. It is literally built inside a titanium mountain, where the ore was mined. The mountain dominated the scene, hundreds of kilometres from any other human habitation, and the workforce would march in and out at the beginning and end of every shift like a river of humanity. It was like stepping into the Land of Mordor in *Lord of the Rings*. Titanium sponge would be worked into blocks of solid metal called billets, then emerge from the mountain on trucks, to be made into the grade 9 alloy that Ain Tec used for our forks.

Unsurprisingly, the people who ran the plant were hard-bitten industry veterans. The feeling of remoteness was inescapable; if you decided you weren't having fun, you couldn't exactly hail a cab and go somewhere else. And the potential for upsets was substantial. We were and are a relatively small customer, and business in Asia works on personal relationships. All of this begins to race uncomfortably through your mind when you're in one of the only restaurants in town, the Chinese engineers are pouring multiple toasts in the local sorghum spirit, and then half a dozen beautiful young women unexpectedly enter the room. Thankfully, they were only there to sing duets with us as the karaoke machine was turned on, although even that was embarrassing enough. If the supplier had been British or European I might have had to walk out – as it was, good relations were maintained at the cost of nothing more than a sore head and a few moderately embarrassing selfies. Part of the art of international business is being sincere and effusive in thanking someone for their hospitality, while simultaneously conveying the message that, in the non-engineering sense of the word, they've gone outside their tolerances.

My first visit to Ain Tec in 2004. Tony Lin (left) and Pierce Huang (next to me) are now my good friends.

When it comes to the actual materials, thankfully there tend to be fewer tolerance issues, because titanium alloys are made that bit more precisely than steels. There's no throwing sacks of charcoal into a cauldron to increase the carbon content like they used to do at Deeley's Castings; it's either the pure metal or alloys with varying and very precise quantities of aluminium and vanadium. The main customers for titanium are also significantly less tolerant of variation than typical bicycle manufacturers; they're in industries like aerospace, medical devices and nuclear waste processing. That tends to mean that, in general, materials tolerances are much finer than in steel tubing. You don't get consignments of tubes that squidge like butter; you either get good parts or nothing at all.

But you can take the best-quality materials, with the finest tolerances, and you can still make something from them that will break unpredictably and injure a customer if it is subjected to stresses that weren't taken into account. Even a titanium frame can be dangerous if it's designed wrongly, particularly if it has some steel parts which might be subjected to entirely different forces if the titanium elements of the frame don't flex in the same way.

Japan in 2008, at a Geisha ozashiki, courtesy of Japanese distributor Toshi Mizutani. Foreign trips can often take you out of your comfort zone!

In order to simulate stresses, we use something called a fatigue rig. This is a machine that attempts to reproduce twenty years' worth of stresses in an afternoon, by overloading a component with multiples of the weight it would normally bear, then bumping it with excess force every few seconds. The original one at the Brompton factory was another of Andrew's custom-made creations, and it was a superb thing; pneumatically powered and utterly precise and reliable itself. During periods when we were making changes to the frame or hinges, it used to haunt me; I would go to sleep with the sound of it ringing in my ears.

Nobody will ever have to go through quite as many iterations of the fatigue rig again, in all probability, because the acceleration of modern computer power means that 'finite element analysis' is now possible on a desktop. A finite element analysis system works in a way not unlike an extremely accurate but very boring video game, modelling a physical object from its design files as a very large (but finite, hence the name) set of polygons, then using accurately simulated computer-game physics to model how stresses applied to any given point affect the deformation of those polygons and the overall performance of the modelled part. Each individual calculation is not particularly difficult, but the complexity builds up because you need to keep track of tens or hundreds of thousands of calculations in order to achieve an acceptable level of realism.

Back in 2002 these systems were the preserve of big car makers or technical universities, as they require huge computer power that was expensive and not affordable for any normal business; these days, you can download a free trial version. But even state-of-the-art software can only work if you start by respecting the problem. If you're not starting with either a few decades of experience of real-world failures or the sort of unique intuitive feel that Andrew had for asking the right questions (and preferably both), then you need to add on several layers of margins of safety just to cover your own ignorance.

This lesson about testing and the real world had been built into Brompton's corporate DNA by a lesson that Andrew learned in the early days. Weirdly, the early problems he had in obtaining financing almost certainly ended up saving the company, because they led to a five-year break between producing the first few hundred bikes that were pre-sold to friends and family, and starting up the factory in earnest. During those five years, quite a few of the earliest Bromptons came back with metal fatigue in unpredictable locations, and they had to be repaired. Andrew was able to process the information coming in from the patterns of fatigue failures and use it to make changes to the design. If he had moved straight from the first run to mass production, there could have been thousands of the bikes out in the wild, and the returns could have bankrupted Brompton.

That kind of knowledge of how bikes perform in use can't be gained in any other way than by having lots of customers and lots of time, but once you have the knowledge, it gives a wonderful feeling of control. Designing something from scratch is nerve-racking precisely because it's so difficult to know what kinds of weaknesses you should be looking for; you don't know what parts to test. When we were designing the titanium frame, though, we had a lot of experience and knowledge of what to look for.

The Superlight was a great seller, and for many of our customers it transformed the usefulness of the Brompton by bringing down the weight to a level they could comfortably carry. But I had always had the ambition to go further and make a full titanium Brompton. To do that, however, we would need a combination of supply continuity and specialist knowledge, and that wasn't going to come from either Russia or China. Whoever was going to make the full titanium frame would be developing huge amounts of know-how and acquiring a level of understanding of the product that felt as if it should be in-house at Brompton rather than in someone else's factory. So we were trying to learn as much as we could ourselves about metallurgy and the mysteries of titanium, and we had started making a few friends in the small but friendly titanium 'community'.

In 2015, I was invited on a trip to the Sheffield University innovation hub, to learn about metal injection moulding, an extremely advanced casting process that is probably more suitable for intricate medical components than for bike frames. While there, I met Steve Kirk, the managing director of a nuclear and aerospace company called CW Fletcher. We talked afterwards, agreeing that metal injection probably wasn't suitable for either of us, but rather than just filing the meeting away as a pleasant evening, I decided to get back in touch and arrange a visit to his factory. I could see that even though CW Fletcher were working on huge assemblies for aero engines, they had the kind of understanding and knowledge that we needed. When I asked Steve if he could help us to find a way to make titanium bicycle frames in Sheffield, he looked at me as if I had gone mad.

The conventional wisdom at the time was that this was economically impossible. Brompton had already looked into buying one of the only 'British' titanium bike producers of the 2010s, but they didn't have the welding capability either. Everyone assumed that titanium was intrinsically expensive to source, manufacture and weld, and so the only way that consumer products could be made out of it was to take advantage of lower labour costs overseas. But as Steve and I talked over the idea, we began to realise that this wasn't necessarily true. A lot of the welding costs, in particular, weren't due to intrinsic properties of the metal; they were related to the way it was typically used.

Most titanium manufacturing involves making parts of aircraft engines, or flasks designed to carry radioactive material, or something equally dramatic. Obviously, these are heavily regulated uses, and consequently very expensive to make. To a large extent, though, what people are paying for is not the actual titanium or even its processing; it's the certification system which accompanies every step of the process. The big difficulty with titanium is that when

A titanium front fork: the Brompton Fletcher joint venture proved that titanium bike parts can be made cost-effectively in the UK.

the metal gets heated up to welding temperatures (you can't braze it), it has a way of drawing oxygen into the crystalline structure and becoming brittle. This, along with the quick-forming oxide layer, is why titanium usually has to be welded in an environment that's flooded with inert gas. And a titanium welder, unlike with most other metals, has to ensure that not only the weld and the liquid surface at the tip of the flame but also the whole metal structure is shielded from contact with oxygen until it's cooled down.

For an aerospace or nuclear application, you therefore need tracking of the materials at every point along their journey and close inspection of the whole piece every time a weld is made. You're also generally working to tolerances much finer than for any bike. The titanium welders at CW Fletcher are very highly trained, and paid accordingly, but a large part of their training is in understanding the compliance requirements they have to meet and the standards required for a piece of engineering to be signed off that really can't be allowed to fail.

A bicycle needs to meet strict safety and reliability standards, but it can't explode or fall out of the sky. It's possible to meet all the relevant safety and reliability standards with a much less regimented certification regime. In fact, you can do more or less what we do with brazers at Brompton – take ambitious young people from an assembly line, test them for fundamental aptitude and then put them through a training process that will leave them able to make frames. Reducing tolerances to the appropriate levels immediately lets you start competing with Russia and China on cost, and once you've added in the benefit of having the jigs designed by bike experts rather than general-purpose titanium engineers, you're more than competitive. Right now, Brompton Fletcher, the joint venture that we set up with CW Fletcher for the purpose, is making some of the best titanium frames in the world, at a cost that's less per unit than when we were commissioning them from 6,000 miles away.

I have made this point before, and it will appear again, very importantly, in the next few chapters as we start moving beyond the manufacturing process itself and thinking about the company as an organisation. The secret is always to keep things simple, and one of the best ways to keep something simple is to reduce the number of different purposes that it has to serve. The Swiss army isn't necessarily the best army in the world, but it's the best army in the world at one specific task – defending Switzerland. Because the Swiss generals know that they're never going to have to fight in the desert or at sea, they can design everything they do around the single thing they know they need to do.

In other words, you design general systems to solve specific problems. Too much of the time, people work the other way round, trying to make multi-purpose processes and then troubleshoot them one at a time. That's how things get complicated. You also have to make sure you're constantly generating the kinds of information that you will need in order to make improvements to the system. That's how I learned to make bicycles, and these sorts of concepts were the basis of how I started applying this kind of thinking to the task of growing a company.

PART 2

BUILDING A COMPANY

The Raleigh Cycle Co. Limited

Nottingham
England NG7 2DD

telephone 0602 77761 - Ext
telex: 37681 (Ralind Nottingham)
telegrams: Ralind Nottingham

21st April 1977

A. W. Ritchie, Esq.,
Director,
Brompton Bicycle Ltd.,
53 Egerton Gardens,
LONDON S W 3

Dear Mr. Ritchie,

Following your visit here on 1st April and the demonstration of your prototype Brompton Bicycle, we have shown the sample only to senior members of the Technical/Marketing and Sales operation of Raleigh Industries.

We have looked at your device very carefully as indeed it has considerable merit. However, it is the concerted view of those senior members that while the device shows much ingenuity it would not be a practical proposition for us to enter into manufacture as it would require a considerable degree of re design; but, perhaps more importantly, that we do not believe that the device could open up a market of sufficiently high volume at the price it would have to be sold at.

The bicycle has already been returned to you at your request and we have your confirmation that it is in your possession.

Thank you very much indeed.

With kind regards,

Yours sincerely,

A. P. OAKLEY
DESIGN DIRECTOR

Building a company is different from building a bicycle. For one thing, a company has a lot more moving parts. And more importantly, many of the moving parts are people; they don't always react in predictable ways, and you can't look up their tolerances in the documentation. Many of the principles are the same, but they have to be adapted to a world in which much less is certain and everything is always changing.

A company is a living, growing entity. And as it grows, it gets more complicated. This is probably the biggest difference between being an engineering innovator and an engineering entrepreneur, two skills which are only very rarely found in the same head. Since I took over the management of Brompton from Andrew, our output has doubled, then doubled again, and is quite near to another doubling. It's just not possible to carry on in the same way at such a different scale – it would be like trying to ride a Tour de France stage as a succession of two-kilometre sprints.

At the same time, some things have to be preserved. The integrity of the product can't be compromised. So, as the saying has it, more or less, everything has to change so that anything can stay the same. As much as I can, I've tried to make those changes gradual and aligned to the growth of the company, so that the overall feeling is of gradual evolution. But that isn't always possible. You can't shift to a new factory one machine at a time. You can't have half the staff working on a production line and half on a craft basis.

So being in charge of the company is not the same as finding out ways to make bikes more efficiently. As chief executive you have to consider different things, like space, people and money. Selling bikes is a different business from making bikes, and developing people is different from installing machines. There are also often times when you have to do things you would rather not, but can't avoid. I had been thinking about all of this since before I joined the company, and I had been taking on more and more management responsibility before I officially became CEO, so I was fortunate to experience the transition to my new role as a gradual change rather than a sudden shock.

In this second part of the book, we're going to focus on the things that I learned about managing a company, rather than on making bicycles. I'll start by talking about money and control. The two go together in my view, because profitability is the basis of keeping control of your own destiny. If the company doesn't make money, then sooner or later you will hand over some form of control to whoever is prepared to finance you. In that sense, making profits is not so much the purpose of the company as the prerequisite for having any other purpose at all. You don't live in order to circulate blood through your veins, but keeping the stuff flowing is an essential part of whatever other projects you have. And it works the other way too; I'll discuss how I became CEO in a management buyout, another exchange of money for control.

Looking at the accounts in some detail will also provide context for a few key events in the history of Brompton as a corporate entity since I took over the top job. The numbers are not difficult to explain; the published results are all available online and the headline figures tell their own story. Brompton Bicycle Limited was set up as a company in 1976, to exploit Andrew's original design. Because of difficulties in manufacturing and getting finance, it didn't make a profit until 1990, a couple of years after Julian Vereker's investment. By the time I joined in the early 2000s, it was a solid small business, with £2 million of revenue and a quarter of a million pounds of profit, selling around 5,000 bikes a year. When I took over the management in 2008, we were selling slightly fewer than 20,000 bikes, with revenue of £6.7 million. Our last reported year at the time of writing, to March 2022, saw more than 93,000 Bromptons sold, despite Brexit and the coronavirus epidemic, and turnover of £109 million. The average revenue per bike has also steadily increased, partly because we have introduced higher value-added options, partly through sales of ancillary gear such as carrying bags, but also because we have taken more control of our distribution network, handling our own exports in a number of key markets and selling directly to consumers. And, of course, each bike out in the world means a stream of spare parts sales, of which we capture a significant proportion because we make so much of the Brompton ourselves.

So the events in this section of the book cover a period during which the company expanded roughly by a factor of ten. When I took over at Brompton, it was with a clear ambition to grow. Our goal has always been to change how people live in cities, to bring

Building a brand is more than just an assemblage of parts.

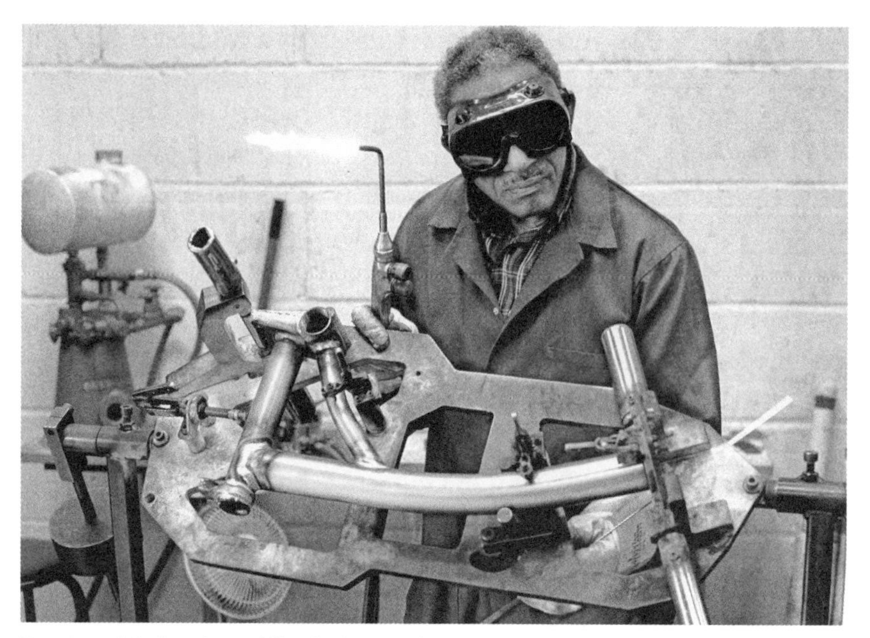

Because of their unique skills, the brazers have always been the highest-paid employees at Brompton. Les Francis at Brentford in 2009.

freedom and to make people a little happier. But we were not delivering this to our full potential: we could not supply enough bikes, and in most of the world no one had heard of us, so how could they be expected to consider having a Brompton in their life? Changing this would be a big project, but one with the obvious first requirement to make a lot more bikes. And this had to be done without making losses or risking the company's finances. The shareholders were Andrew's friends, my own friends and the employees; people who had backed the vision from the start and who deserved to be looked after. Growing the company has been the most important task for more than ten years, and in order to do it we have had to completely reorganise the production process, move to a larger factory, and change the way we sell and distribute the bikes. In fact, the factory move is worth considering from several angles – not only why we needed to do it and what needed to change, but also the mechanics and organisation of how we went about it. Moving to the current factory in Greenford was a big physical and organisational change, but it also involved some big financial and managerial changes.

And then there's the people. In many ways, a job in management is defined by the things that you have to trust others to look after. If you know everything that's going on, then you're not managing anything – it's you that's doing the work. Everything I've learned by applying engineering principles to manufacturing industry also applies to this fundamental fact of management.

But I'll save some of the upbeat stuff for later. Let's look now at some of the most difficult experiences I've had. It is not easy for anyone to be the second chief executive of a company founded by a genius. 'Founder's syndrome' is well known: the state of affairs in which someone has passed on official control, but is still there and still expects to be consulted and to make decisions. I began my time at the top of Brompton by making some decisions which I regarded – and still regard – as completely necessary, but they were not pleasant, they caused quite a lot of bad feeling and Andrew fairly unreservedly disagreed with them.

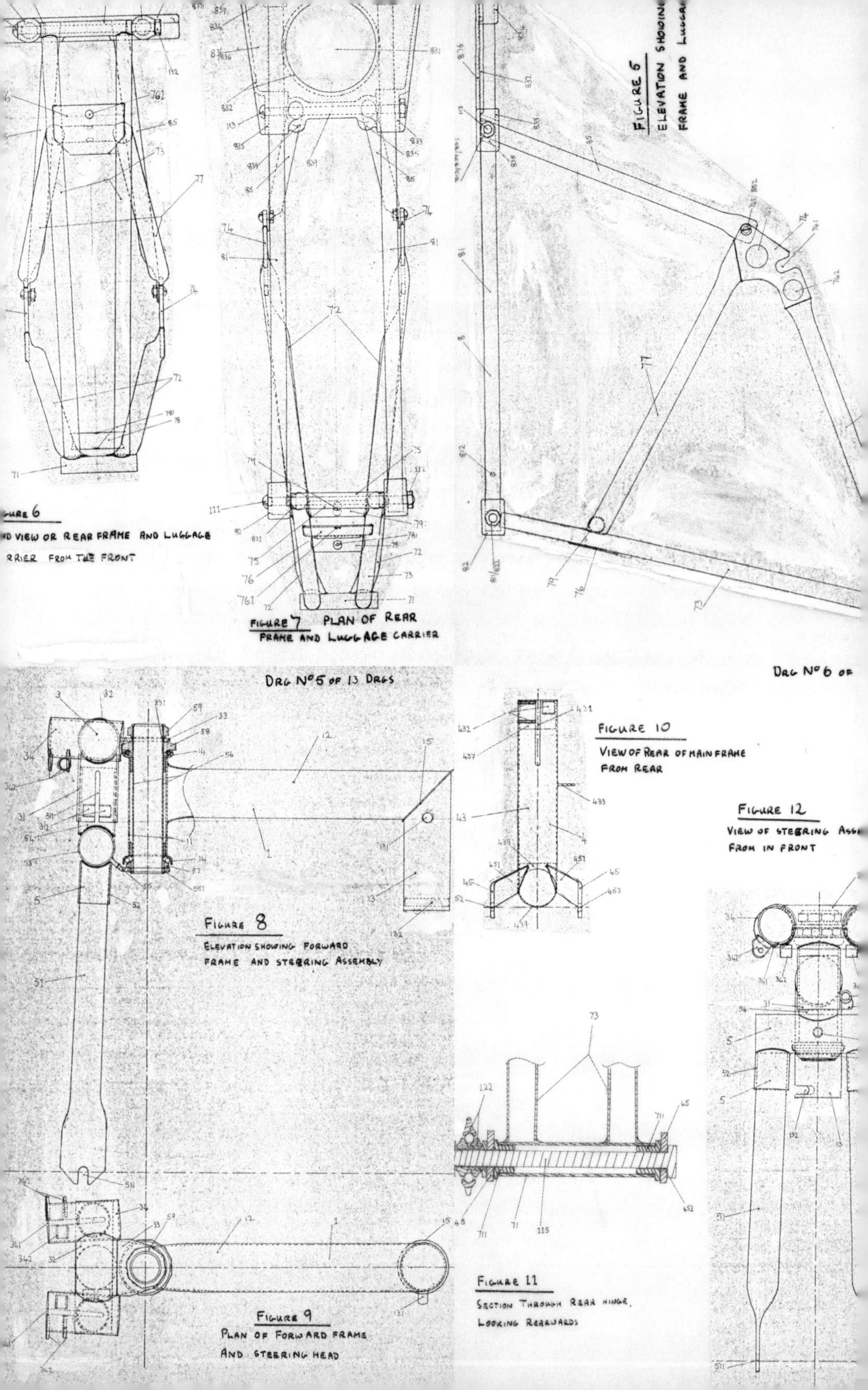

FIGURE 7 PLAN OF REAR FRAME AND LUGGAGE CARRIER
FIGURE 5
DRG Nº5 OF 13 DRGS
DRG Nº6 OF
FIGURE 8
ELEVATION SHOWING FORWARD FRAME AND STEERING ASSEMBLY
FIGURE 9
PLAN OF FORWARD FRAME AND STEERING HEAD
FIGURE 10
VIEW OF REAR OF MAIN FRAME FROM REAR
FIGURE 11
SECTION THROUGH REAR HINGE, LOOKING REARWARDS
FIGURE 12
VIEW OF STEERING
FROM IN FRONT

6 LINES AND BATTLES

> *When the UNITE trade union was invited in to organise the workforce, their representative showed up riding a Brompton. This well-intentioned gesture backfired badly; the union had looked at the contracts and thought they were fair, but a rumour went around the factory that management had bribed the rep with a bike.*

It was a kind of poetry in motion watching Calvin work. One of our best assemblers, he would take one glance at the pink sheet which specified the options on a bicycle and move like a dancer, picking the right component every time and fitting it without a wasted movement. Picking up the frame and setting it down again at a new angle, over and over again without tiring, completing the job and then on to the next one in a mesmerising flow state. He could build twenty, even thirty Bromptons in a shift; some of our other workers managed seven.

How do you tell someone, after they have brought their craft to that level of mastery, that their skills aren't useful any more?

Making something ten times bigger changes things. You can't just do ten times more of the same, because you would be scaling up all the problems and inefficiencies too. Something that's merely an irritation at small size can become a show-stopping problem at scale. Unfortunately, that means change, and change means conflict. People are invested in the way that they do things. This can be obvious when they've developed particular skills that are adapted to one version of the process. It can be less obvious, though. The early experiences of Brompton as a company shaped its culture. As far as I possibly can, I've tried to keep the positive aspects of that culture and to emphasise flexibility and openness. But like a human being, a company's personality is partly defined by the things that it doesn't like and wants to avoid.

The whole point of keeping employee turnover down is that people get experienced and they get good. At the whole company level, there has to be some sense of identity and coherence. But these are also the forces that make change difficult. Part of being in charge is that you have to face up to this from time to time; the CEO occasionally has to push the system out of equilibrium so that it can adapt to different conditions. But it's not the nice part of the job. When I was promoted, I knew that I had to change the whole system at Brompton, and that it wasn't going to be easy.

The underlying reason why Brompton was pushing up against capacity constraints was that there was no flexibility in the operating model. What does that mean? Well, consider how Brompton Bicycle Limited is affected by the progress of the earth around the sun. Bromptons aren't toys, so they don't have the same Christmas demand as recreational kids' bikes. They're generally bought in the summer, when the idea of commuting to work by bike seems more attractive and crowded public transport is at its most hellish. Although there are plenty of year-round riders, it's an unusual person that wakes up in the dark, steps out into driving sleet and rain and thinks about buying a folding bike for the first time. This makes demand for Bromptons quite seasonal.

On the other hand, the potential output from the factory is more or less constant on any given working day. If it wasn't regulated, we would either pile up excess bikes in the winter, find ourselves short and unable to fill orders in the summer, or (as actually happened in the early days) a bit of both. And producing bikes for stock is incredibly inefficient. It's financially bad news, because it lengthens the gap

New assembly line at the current Greenford factory in 2016.

between incurring the costs and getting the revenues, meaning that you have your cash tied up. And it's commercially bad news: if you guess in November what styles and colours are going to sell in May, that guess will almost always turn out wrong. Shortages and waiting lists slightly mitigated the problem – when distributors were conditioned to expect Bromptons to be in short supply all the time, they bought when they could get them, moving some stock off our balance sheet and onto theirs. But that's hardly a way of managing things, and it even caused its own problems. More than once, we found we had accidentally become 'channel-stuffed': we couldn't sell as many bikes as we wanted because distributors had previously over-ordered.

The solution is clearly to make the factory output more flexible. In the early days, this was practically impossible to achieve. In our romantic but somewhat outdated model, each bike was made by a single craftsperson. Just as the brazers' stamps were an individual mark of pride as well as a quality control feature, in principle you could come to the factory and meet the 'assembler' who made your Brompton. The assemblers were practically as high-status as the brazers or the inspectors. They did a highly skilled job which took a lot of training. They were also on piecework, so if they were skilled and athletic enough they could earn multiples of the pay that we'd anticipated during a given period. They were supplied with large parts (handlebar assemblies, rear frames and so on) from the 'pre-assemblers', and there were also 'pre-pre-assemblers' who made the gears, chain pushers and other small components that we produced ourselves instead of buying in from Taiwan.

In a way, this creates the same problem as a Rubik's Cube – combinations multiply, and the numbers involved get out of control really quickly. The assembly manual, a huge document which Andrew had originally put together, broke the job down into stages. The details changed over time as new models and designs were adopted, but typically you might have allowed an hour to make the bike, broken up into twenty tasks, each taking an average of three minutes, and each of which required the assembler to choose between maybe five different options.

And here you have the difference between craft production and assembly-line production. For any one of those stages on an assembly line, someone might have to learn a single task with five options. You could put the parts required right next to their work station, with their tools next to where they would be needed, ergonomically designed. In craft production, on the other hand, if one person carries out all twenty tasks, the possibilities multiply up – in fact, five to the power of twenty equals more than 9 trillion. It's a testament to the versatility and adaptability of the human mind that this is possible, but it takes training and experience. The assemblers were on a level with the brazers in the old factory precisely because not everyone could do what they do. And as with the brazing shop, the capacity of the assembly shop was more or less fixed. We could do our best, introducing flexible working practices and encouraging everyone to take their summer holidays in September rather than May, but the seasonal demand problem was more or less insoluble.

The cause and effect here isn't difficult to see, but there are a few subtleties to it. You could look at the factory and say that the assembly stations were a bottleneck, but it's worse than that – the whole organisation was a bottleneck producing machine. Every time we added an assembly station, we needed to add an entire new set of tools and arrange for the station to be supplied with all the components for every permutation of options. This meant more space and more complexity; not only was output seasonally inflexible, but it was not geared to long-term growth either. Nigel Saffery was one of my key recruits as lean manufacturing manager, and we had been talking about this problem ever since he joined in 2006. Nigel had previously been in charge of a factory reorganisation at Desoutter Tools, before doing a masters degree at the University of Hertfordshire and becoming an expert in lean manufacturing. It almost became a slogan between the two of us; everything was judged on the basis of 'What is this going to look like when we're making 50,000 bikes?'

It's worth noting that the assembly bottleneck had a different cause from the brazing bottleneck, which made it a different problem. Brazers took years to train because the skill was intrinsically difficult and people had to develop muscle memory, understanding and concentration skills. Assemblers took years to train because they had to learn a lot of combinations. It was a complex task rather than a difficult one: the two words are not synonymous. And of course,

Will Carleysmith (at left) and Nigel Saffery at our first Brompton World Championship in Barcelona in 2006. You need to share the vision to work at this company!

the solution was found by Henry Ford when he kick-started the modern industrial era – you break complex tasks down and create a production line. It's a lot easier to solve twenty single-cube puzzles than a whole Rubik's Cube. Working this way, we would be able to recruit temporary workers every summer – it didn't matter if they were students trying to earn some extra money, as they would only have to master two or three relatively simple tasks. Obvious?

If only it were. There are two main reasons why companies stay with craft production rather than moving to a line model, and they can be powerful incentives not to change. Have a look at a software development house, for example, and see how many bottlenecks and delays can be traced back to the fact that key components in the finished product are dependent on the idiosyncratic skills of a single programmer who is the only one who understands how to put things together. The two reasons why managers tolerate this sort of situation are, first, they don't think the system can handle the conflict that change causes, and second, they don't know how to break the tasks down into smaller units.

Breaking down tasks is difficult, and putting them in the right order is also surprisingly tricky. The best sequence for a production line isn't necessarily the logical order in which you would build an individual bike – you might work from the outside in, or assemble two parts at different stages even if they will end up physically right next to each other. The key principle is for the line to be 'balanced', so that it gets as close as possible to the ideal model of twenty tasks, each taking three minutes and each carried out at an ergonomically optimal station. And that's a fundamentally different mode of thinking from craft production.

To get an idea of the shift in perspective that's involved, think about an underground station escalator. There's a really paradoxical phenomenon known to public transport experts, which is that the quickest way to get people in and out of the station is to have them standing on the escalator in two lines. Yes, anyone who would otherwise have walked up or down will now be moving slower and will take more time, but because the standing lines use the space on the escalator more efficiently, the total number of bodies moved per minute will be higher. (Try this on your colleagues; about a quarter of people are unable to see that the two measures can be different.) Something similar applies to production; a speedy individual operation is only a benefit if it doesn't lead to part-finished bicycles piling up in front of the next station.

This is particularly visible in the early days of any new production line system, or whenever significant changes are being made. You tend to find that the initial chunking of the jobs was a bit off, so you may have a couple of two-minute tasks feeding into one that's closer to five minutes. Or more problematically, there's a task which is more complicated for some configurations than others, so it could take two minutes or it could take six. The way to deal with this is to start with bigger chunks to begin with, then break them down further as you learn. Balancing the line also needs to be a dynamic activity, to take account of variable configurations.

Nigel had some experience with doing this from his time at Desoutter. We sat down together and started to work out how the tasks could be divided up and spatially organised. But during the initial stages, the line still regularly got out of balance and ground to a halt.

That's actually one of the good things about the line system – if you have a problem on a production line, it can't be hidden or mitigated by someone cutting a corner or hurrying up on a later stage.

That's valuable, because it generates exactly the kind of information that you need to keep incrementally improving productivity. Once more, and at a higher level of organisation, we've rediscovered one of the things we learned from making bikes. The key to making things work is to try to turn all the production systems into information systems, so that the organisation itself generates the capacity to solve problems as it uncovers them.

But this benefit of the production line can a bit theoretical and insubstantial when you're trying to explain it to someone who was sceptical about the whole thing to begin with, and is staring pointedly at a factory floor full of people mostly standing around doing nothing, except for one station where a group is clustered around the poor soul whose backlog is holding everything up, giving him helpful comments and advice.

I'm not sure what the point was of Andrew bringing the hold-ups to my attention; it didn't cheer me up and it didn't calm him down. By that point, the floor had already been reorganised and it would not have been practical to move everything around again in order to go back to providing a dozen individual workshops. But Andrew wasn't the only one to be upset, and some of the other disagreements around the factory were based on fundamentally opposed interests, not philosophical preferences about manufacturing.

One major problem was that any change to the production line was a direct attack on the piecework system. That change was extremely unpopular, but what surprised me most was how strongly Andrew opposed it. I knew that to some extent he liked the idea of each Brompton being made by a single person, but one wouldn't expect a mind like his to rate the romance of craft production over the efficiency of an assembly line. While Nigel and I worked to overcome Andrew's objections, we learned a lot about his true reasons for them, and then we learned even more from his complaints once we had finally made the changes.

The basic problem Andrew had with the production line system was that he was, at heart, an individualist. The way things worked in his view was that a person made a bike. If they were good at it – smart, dextrous and athletic – then the piecework system would ensure that they got paid well, and there was no limit to how much a really talented employee could make. He was right: by the time Brompton was moving over to the production line system, several brazers and assemblers were earning more than I was as managing director or Andrew was as the founder and owner, and he was totally relaxed about that.

The line, on the other hand, is intrinsically collectivist. It moves at the speed of the slowest station, and if someone is a much faster worker than his or her colleagues, this just means that they spend a few minutes doing nothing and waiting for the previous station to catch up. There's no individual responsibility and nowhere for the inspectors to send things back to. This is not a bad thing, or at least, it wasn't an accidental consequence of the reorganisation. The point of the new system was that the responsibility is collective. We started to look at the concept of 'skills-based pay' to replace piecework. Brazers would be paid in relation to the number of operations and joints they were qualified for, with a premium for being able to work to raw lacquer standard. Assemblers' pay would be based on

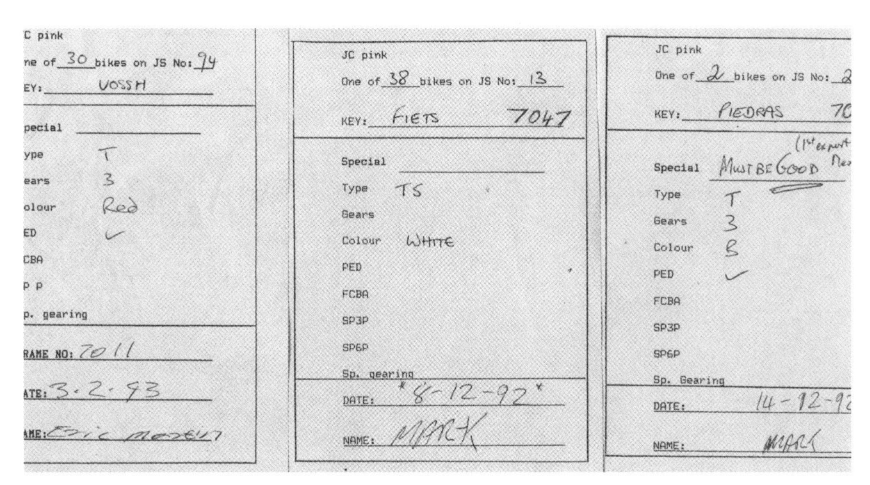
C pink
ne of 30 bikes on JS No: 74
EY: VOSSH
pecial
ype T
ears 3
olour Red
ED ✓
CBA
P P
p. gearing
RAME NO: 7011
ATE: 3-2-93
AME:

JC pink
One of 38 bikes on JS No: 13
KEY: FIETS 7047
Special
Type TS
Gears
Colour WHITE
PED
FCBA
SP3P
SP6P
Sp. gearing
DATE: *8-12-92*
NAME: MARK

JC pink
One of 2 bikes on JS No: 2
KEY: PIEDRAS 70
Special MUST BE GOOD (1st export
Type T
Gears 3
Colour B
PED ✓
FCBA
SP3P
SP6P
Sp. Gearing
DATE: 14-12-9
NAME: MARK

The 'pinks' – slips of paper which set out the model configuration of each order for the assemblers.

how many different stations on the line they worked at. Effectively, this is modelling the same philosophical change; we now reward people for how useful they can be to the factory and company as a whole, rather than for individual efforts which don't necessarily translate to the flow of bicycles through the system.

But although a move from piecework to skills-based pay would mean 80 per cent of our workers seeing a moderate increase, in order to hold the overall pay bill constant (which we did) the remaining 20 per cent were going to see a very noticeable cut. For the brazers and assemblers who had been doing best out of the piecework system, this cut might be as much as 20 or even 30 per cent of their income. Brompton had a plan to soften this and avoid making anyone cope with such a huge reduction to their family income all at once. Another part of the plan was to introduce an annual cost-of-living increase for the first time. We proposed to freeze the incomes of the top producers in nominal terms, until these annual increments meant that the skills-based levels caught up with them.

Jesus Blanco, one of our best assemblers, could build twenty or more Bromptons in a day: Brentford, 2006.

The electric Brompton took nearly as long to bring to market as Andrew Ritchie's original design.

Assembly takes place at individual stations in the Greenford factory, one per task, with the line arranged in logical order to keep timings balanced.

A part is held in place by a jig in order to be brazed.

A perfectly brazed joint should be 'linear', with the metal flowing in regular 'stitches'.

The automatic brazing machine handles the most complex joints, with a thermal imaging camera to measure the temperature.

HRH the Duke of Edinburgh during his visit to the Brentford factory in 2010, with head brazer Abdul El-Saidi (at left) and Andrew Ritchie (centre).

The additional space at Greenford allows the factory to be laid out efficiently with good ergonomics and workflow.

One of the very first batch of 50 pre-production Bromptons made by Andrew Ritchie in 1981.

The first production run of 400 Mk1 Bromptons (this example is from 1982) had a more pronounced 'hump' in the frame than later models due to the tube bending process.

This Mk2 Brompton from 1996 has the classic shape which has remained as the visual signature of the brand.

The heart of the Brompton design is the hinge.

After casting, the hinges are machined to extremely precise dimensions in a computer controlled cutting tool.

A parts bin holds hinges close to where they are needed ...

... and they are brazed onto the steel frame tubes.

Casting malleable whiteheart iron for Brompton hinges.

Welding titanium components in an argon atmosphere.

Different configurations are available to suit different riding styles.

The Brompton needs to be as light as possible, so that it can be carried as well as ridden.

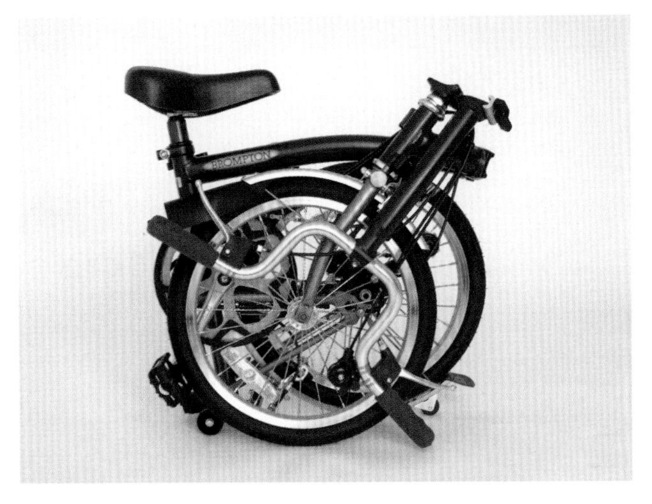

The central hinge allows the bicycle to be folded into a compact package with all the oiled parts kept harmlessly in the centre.

This mural decorated the mezzanine at the previous factory in Brentford.

The new Greenford factory covers 100,000 square feet.

Space and light turn a factory floor into a place where people can feel proud to work.

Spokes are fed into the hubs by hand and then 'laced' and 'trued' by machine.

The Holland Mechanics automated wheel-lacing machine uses laser sensors and state-of-the-art robotics.

Brompton bicycles are painted by an electrostatic process, in which the parts are coated in plastic powder which is then heat-cured.

Frames need to be handled carefully through the paint shop to avoid 'inclusions' and other small but visible defects.

There are six standard colours, plus special editions and a transparent 'raw lacquer' finish which allows the brazing to be seen.

The Brompton Junction shops were developed to show the bike to its best effect. By now there are sixteen Junctions, with flagships in London, Paris and New York.

The Brompton Junction in Beijing, China, in 2015.

Below: The New York Junction.

Above: The first Brompton Junction opened in Kobe, Japan, in 2011.

Me with a dazzlingly customised Brompton and its owner, Singapore, 2013.

Tony Huang and family on the Old Caoling Tunnel bike path, on Taiwan's north coast.

Ruth Nicol, Mark Woollard and Andrew Barnett ('Bumble Bee') during the Brompton Urban Challenge 2014. 'This was taken by me, Tim Eavis, at Waterloo Station, London. Mark, Bee and I founded the London Brompton Club, and we won the Challenge that year.'

BBC Radio presenter Sara Cox on her electric Brompton, 2018.

Berenda Lim catches the sunrise near Mount Fuji, Japan.

Serena White's 2009 M3R in flamingo pink moved with her from London to Sydney, Australia, in 2013. 'Cycling to work across the Sydney Harbour Bridge is awesome!'

Brompton fans come from all over the world to race in the Brompton World Championships. The factory team is often highly placed, but has yet to win the 'best dressed' trophy.

RACE

David Millar, former cycling champion, helped design the popular CHPT3 special edition.

The CHPT3 has titanium parts, BMX-style handlebars and appeals to cycling enthusiasts.

The copper plating on the 'Tom Dixon' special edition created massive technical challenges, but the lessons learned were valuable for the future.

Of course this was only a slight cushion to the announcement that the golden days of piecework pay were over at Brompton. The people who were going to lose out from the change still hated it. And it wasn't just a matter of democracy, with the 80 per cent who were going to gain from it being in the plan's favour. By design, the people who lost out under the new system were the best employees we had; the most skilled brazers, the quickest assemblers, the people who could build twenty bikes in a day. They were used to being the stars, who everyone deferred to and wanted to learn from – they were the ones that I looked to myself in order to understand how to make improvements. For most of Brompton's previous history, the problem had been not being able to make enough bicycles, and these were the employees who thought they were doing the most to help solve it.

Effectively, there would be no more master assemblers; everyone was doing tasks of the sort which would have previously been called 'pre-assembly'. And given that the line had to move at a constant

The incentives built into the piecework system meant that assemblers always wanted the fastest jobs.

speed, there was no reason to incentivise anyone on it to work faster. The days of the dancing martial arts masters of the assembly stations were over. But unfortunately there was no easy answer to the question at the start of this chapter; because there is no good way to tell someone that the skills which made them one of the most special people in the factory don't have a place in the system any more.

At this point, it was almost as if the factory had an immune system and it had identified a disease; the production line defined the new arrangements as alien and a threat and reacted accordingly. From being the smiley face of Brompton's management, the friendly guy who people could talk to if Andrew had been sharp with them, overnight I became the villain, the heartless manager who was tearing apart the community that Andrew had built. Between 2007 and 2010, we had to go to court three times for employment tribunals as people started accusing the company of unfair dismissal or constructive dismissal (when an employer makes your job so intolerable that they might as well have fired you). Tragically, in most cases (including those of the people who just resigned), the employees who left genuinely loved the company. They had just fallen a bit too much in love with a particular stage of the company's life cycle and weren't able to move on when things had to change. Or to put it more harshly, perhaps they had fallen in love with those aspects of Brompton Bicycle, vintage 2003 to 2004, which most closely reflected their own talents and aspirations.

Common sense tells you that if you're taken to an employment tribunal, you settle. Court hearings are expensive, time-consuming and emotionally draining performances, and almost nobody comes out of them feeling like a winner. But I went through all of ours, didn't settle a single one and 'won' all three. If these tribunals had just cropped up in the ordinary course of things, the temptation to take the easy way out would have been substantial, but this was, with very little exaggeration, a fight for the soul of the company. It was certainly a battle for the validity of my own role as the newly promoted boss. Although the specifics of each case varied, I think everyone understood that the disputes had the same basic cause: the pay system and the assembly line. If I didn't show up to fight for them at the tribunals, I would have been admitting that I didn't fully believe in them. And that would have destroyed the whole plan. The only way to carry it out was in the firm and honest conviction that this was the right way to run the company.

I needed to present my case outside the court as well, and keep making the underlying argument for the plan – it's not just a matter of people being too stubborn to face facts. There's a real choice here, and after all, I didn't really manage to convince Andrew Ritchie, one of the cleverest people I know and someone who cared about Brompton Bicycle Limited so much that it hurt. Let's reiterate the key points here, so that they don't get lost in the story. In the last chapter we talked about growth and complexity, but in this one we've been looking beyond that to management's higher-level responsibilities of flexibility and information. And as we saw earlier, information and quality are intrinsically linked in an engineering sense, because they're both different aspects of the overall problem of tolerances.

It's probably obvious that you can't make 5 million bikes a year on a craft basis with individual assemblers working on their own stations. It's equally obvious that you can't make a single production run of 500 bicycles on an assembly line. But what about the scale at which Brompton operates? It's not completely impossible to make 50,000 bikes a year on the old system – if one assembler could make ten a day (somewhat slower than the rated output written into the old piecework system), then over 250 working days you would only need twenty assemblers. Make some allowances for holidays and sickness, and you might need to train thirty people and kit out two dozen stations. So the people arguing against the new system at every step weren't living in cloud cuckoo land – it would certainly be physically possible to have a world in which every Brompton was still assembled by a named person you could come and shake hands with.

But mere physical possibility isn't the problem here. We started off by thinking about the problems Brompton had in coping with even very predictable seasonal fluctuations in demand. The stocks were only one of the ways in which cash would be sucked out of the business, though. If you have two dozen assembly stations, then even if you ignore the very large factory space you would have to rent to put them in, every investment you make has to be multiplied two dozen times. For example, there are some stages of the process where things have to be fixed on with bolts. For this, we use a modern torque wrench, with a motor and sensors in it to tighten the bolts to exactly the right torque. Depending on how sensitive and complicated it is, one of these tools can cost well into five figures. On the assembly line, you need one of them, and it's in use throughout the day. On a craft model, we might have needed to buy fourteen or fifteen, each of which would be used for possibly three minutes in the course of an hour.

Cash that's tied up in stock or in duplicate tools can't be spent on other things, like research and development to improve the bike or invent new products. Even if cash wasn't a constraint, there's not much incentive to invest in semi-automation to improve efficiency if the piecework system means that the largest chunk of the return on that investment goes straight into the pocket of whichever assemblers learned to use the new technology the quickest. And now, every improvement we make affects the whole factory, not a single station. So the first big reason for changing the factory is that it sets the system up to change again in the future as it grows. The physical signatures of our philosophy these days are the wheels on the parts trolleys, the fixing points on the factory floor, and the overhead power supply, all showing that we can move anything around if we need to.

But the other fundamental point is that the 'collective responsibility' of the assembly line system needs to be physically built into the production. Making that happen was as big a cultural change as the abolition of the piecework system, and just as controversial.

Back when the master assemblers practised their art, the inspectors were on a par with them and the brazers, making up the three branches of the nobility of Brompton. They passed or failed completed bicycles according to a huge 'bible' of standards which had been set by Andrew himself. If a bike took an hour to build, it might take half an hour to inspect and could be sent back for the assembler to correct defects 'in their own time', so reducing their effective hourly rate. Obviously this was a constant source of human resources conflicts and real or imagined personal grudges, but the problem was deeper than that.

In a small, craft-level production system, you ensure high quality by inspecting defects out, and you have to wear these kinds of problems. A big part of the point of reorganising the line was to make sure that quality was 'designed in'. All of the pokey-yokeys, gauges and interim stages that we were talking about earlier – the purpose of these is to create a situation where the process is consistent at every individual stage. Since the final product is the sum of the individual stages, once you have set the line up with the constituent checks, nearly all of the inspection work has been done by the time the bike is complete. This is the way in which you can get growth to start working in your favour rather than against you. Rather than tolerating the normal tendency of growth to generate complexity and proliferate problems, you design the system so that it generates information at the same rate that it produces chaos.

1. **General report** setting out the facts relevant to commercial exploitation of the Brompton Bicycle.

   The market, production requirements, strategies for production and proposals for sale of rights or joint venture agreement are dealt with.

2. **Appendices.** 1. The Market - U.K. and overseas sales
   a) The Market – U.K. and overseas sales figures.
   b) Remaining development work.
   c) Tooling and jigging requirements.
   d) Costings per bicycle.
   e) Assumptions and calculations for proposed strategies.

3. **Patent specification**

4. **Technical information,** drawings etc.

5. **Specific proposals** for sale of rights and joint venture scheme.

For further information
please contact:

A. W. Ritchie,
Brompton Bicycle Limited,
53 Egerton Gardens,
London SW3

Telephone: 01-581 2282

7 THE REAL MEANING OF MONEY

Will joined the board of Brompton in 2006, four years after being recruited. In 2008, he led a management buyout of the company, investing half a million pounds to buy part of Andrew's stake.

It started off as a bit of a cough, but it got onto my chest and felt like it was never going to leave. I was working far too hard at the time, doing a whole day in the factory, then spending every evening going to two or three meetings with investors. The sensible thing to do would have been to take a few days off and recover. But this was the most crucial point of the management buyout, when I needed to persuade new people to put money into the company and existing shareholders to back the deal that would leave me as the CEO and see Andrew stepping back from operational control. So I kept going, looking and sounding progressively worse and worse. I struggled home at the end of one long day, and my wife took one look at me. Three hours later, I was being gently scolded in the accident and emergency department of High Wycombe hospital with a case of pneumonia.

This was all driven by what I needed to do with the company. I had initially agreed to work for Brompton on the understanding that if things worked out, I'd be promoted. While Tim Guinness was recruiting me in 2002, we had discussed the future, and he'd admitted it was possible that I could develop far enough to run the company. This wasn't something that Andrew had anticipated to begin with, but after four years in the job, I'd outgrown my original brief of 'new projects'. I was starting to understand that the company needed to grow in order to realise its potential. And in order to grow, it needed to change things – to reorganise the line, introduce different working practices and modernise the space. If I was going to make that happen, I needed to be in charge. And that meant that I had to get involved with money.

In business, money represents lots of things, but the most important one is control over your own destiny. Like quality and information, tolerance and uncertainty, you can't separate the two facets. If your company makes a profit, then you can decide what you're going to do next. If the profit is enough to cover the investments you want to make, you're fully independent; grow the business, try something new or just pay out a dividend, whatever you like. But if you're not in that position, you are going to be giving away a greater or lesser degree of control to someone else. Borrow from a bank, and you now have an entirely new set of priorities, because the first claim on your next year's operating income is the interest bill. Sell shares and you are literally selling voting rights; this might seem harmless enough at the time, but you're potentially giving away a veto on some future decision. If you bring in venture capital or private equity money, then you are often explicitly agreeing to a strategic plan driven by the priorities of their fund investors, people you will never meet and who have little interest in what you want to achieve.

Of course, not every company is going to be able to avoid taking on debt or selling equity; Brompton was fortunate in being able to cover its investment costs by selling bikes. The worst possible thing , though, is to find yourself accidentally getting into a situation where control slips away from you. That could happen because your accounting system and business plan aren't based on reality, or because you yourself are in denial about the intrinsic profitability of your company. I've seen too many people stuck in companies that are slightly less or slightly more than breaking even, never really

BROMPTON BICYCLE LTD 2

Date			Details	Ref	Debit	Credit
1980						
24	3		Telephone deposit paid by BBL for BB(S)L	1L		50 00
30	04		Cash drawn from BBL bank acct – loan	PCB1		250 00
9	06		" " "	PCB 2		200 00
16	06		Cash rec'd by BBL from G. Philipps, as loan to BB(S)L, wrongly paid to BBL	5G	250 00	
16	06		Cash rec'd in error by BBL from Lady Watts, as loan to BB(S)L	5H	250 00	
19	6		Cash drawn from BBL bank account – loan	PCB 2		50 00
24	6		Loan from BBL, being payment made on behalf AWR	1₃		104 17
4	7		Cash drawn from BBL bank acct – loan	PCB 2		70 00
11	07		" " "	PCB2		50 00
21	07		" " "	PCB2		50 00
29	07		" "	PCB 2		50 00
5	08		" "	PCB2		40 00
5	12		Cash paid by BB(S)L on behalf of BBL – copying AGM	PCB4	7 83	
1981						
13	01		Cheque paid by BB(S)L on behalf of BBL – Cornelius & Co	CB3	102 50	
18	02		Payment by BBL of debt to Nylonic on behalf BB(S)L	6A₄		7 37
9	03		Cash paid by BB(S)L on behalf of BBL – patents etc	PCB6	14 39	
31	03		Balance C/D	c/d	429,19	✓
Mar 31				[illegible] 296.82 [illegible]		
Mar 31			To By Journal	1	2 00	
..			By Journal	1		134 37
					1,033 97	1,033 97
1	04		Balance B/D			429 19
1	05		Cheque paid by BB(S)L to BBL	CB4	75 00	
27	07		Payment made by BB(S)L on behalf BBL, Registrar Co's (for BBL's annual return)	CB6	20 00	
30	12		Payment from BBL to BB(S)L	CB11		3000 –
15	1	1982	" BBL deposit a/c " –	CB11		1300 –
1	2		Transfer from BBL current to BB(S)L	CB11		1000
26	06	81	Bank error BBL charged instead of BB(S)L — Journal	24.25 on Journal 3		141 31
31	03	82	Balance C/D	C/D	5775 50	
Y/E 31.03.82					5870 50	5870 50
1982						
01	04	82	Balance	B/D		5775 50
27	07	'82	Transfer (Bernard Davis payment)	42C	669 30	
30	09	82	Transfer from BBL deposit	CB23		2000 –
10	01	83	Cheque paid by BB(S)L on behalf BBL (B[illegible])	CB26	230 –	
10	01	83	" " " " " (Burn [illegible])	CB26	264 50	
30	11	82	VAT return – transfers to BBL	11B 23/24	594 18	367 99
31	12	82	" "	11B 30/31	285 95	123 44
31	12		Balance c/d			
					2043 93	8266 93

Accounts were all meticulously kept by hand by Andrew Ritchie in the earliest days.

making enough to do what they want to do, and still thinking that there's some sense in which they are their own boss.

Some people learn how to understand the financial implications of a company's business model at business schools. I was luckier; at the age of thirty-three, I got the kind of education that people pay thousands of pounds for, and I got it for free. The course was called 'How to make a management buyout bid for Brompton Bicycle Limited', and my generous educators were the bankers employed by the investment office of one of Britain's wealthiest families. A bit more than a year before my hotel presentation, someone I knew via a family connection had got in touch informally and suggested that this investment office thought Brompton had a marvellous product but could do a lot more to realise its potential. Who was I to disagree? Julian Vereker had said the same thing two decades earlier.

Board meetings were pretty informal at Brompton at the time; there wasn't much of an agenda, and things often rambled somewhat. As things were winding down after one meeting I casually mentioned that I had been approached and asked if I should do anything about it. I was told to find out how much they were willing to pay. But of course you can't get an answer to a question like that just by asking. Management buyout investors need a lot of detailed information, and often have strong views about the future direction of a company when they buy it, including making changes to the board. It was my job to take the minutes at our meetings, and I made sure it was clearly recorded that I had been approved to do what was necessary.

I spent three months talking the bankers through the business, taking them round the factory, explaining how distribution worked, and everything else from Japanese VAT to the price of titanium. In return, they showed me how to value the company. Or to put it more precisely, they showed me how to not undervalue the company, or my own contribution to it. In everyday life, modesty is an attractive personal characteristic, but there comes a point at which you have to assign numbers to things. Going through the exercise of setting out a fair incentive package for myself, the key staff and future management really opened my eyes to this. I ended up concluding that a lot of what bankers get paid for is for facing up to things that normal people find embarrassing and awkward to talk about. One of the most awkward aspects of the deal that the investors wanted to propose would have been that it didn't include an ongoing role for Andrew, and also that it envisaged Tim being replaced on the board.

The deal didn't happen, as there was no way that Andrew and Tim would have accepted being turfed out. I had expected this. But it acted as a catalyst for change by setting out an independent valuation and incentive package, which I was then able to use. The only difference was that I would have to raise the cash myself rather than get it from a family office. Within the next year, I had started to make my own proposal to the board, making the case that the company needed to expand, that Andrew needed to stop overworking and start enjoying some of the wealth he had created, and that I could take on the task of making the big changes that would follow. I used the previous offer as an exact template, but with the ambition of raising the money from a group of people rather than a single bank or investor, so as not to give up so much control. The amount involved turned out to be £2.5 million – not a small sum for a company which was not particularly well known at the time.

The presentations I made at the time were the usual kind of thing you see in five-year corporate plans, with bars constantly increasing and pie charts showing the inexorable progress of export markets. Of course, they bore no more resemblance to reality than any other five-year plan, but that's actually OK. The purpose of making forecasts is really so that you know what to be surprised by, and the important thing about a plan is having the agility to change it. Instead of going through all the projections and out-turns, what I'll do now is give a sort of mental model of how the business is represented in the financial accounts, and then consider how the nature of that task means the business is always misrepresented.

The numbers in the accounts have the big advantage that they are, as much as possible, consistent – the principle of double entry means that everything comes from somewhere and everything goes somewhere. The corresponding disadvantage is that you lose the perspective which tells you the reasons *why* things happened. The reality of the business has to be summarised in a few sentences from the management trying to explain why a few numbers changed, when those numbers are really the aggregate outcome of dozens or hundreds of underlying business events. They're best used as a tool of control, relating to the business plan in the same way that the instrument panel on an aircraft relates to the flight plan – it tells you when you're heading off course or in danger of flying into something.

Taking all these caveats into account, it makes sense to start by taking a very big-picture view, over a number of years, and then zooming in to look at the detail. I'm now going to look at an artificial statistical construct which we might call the 'ten-year average Brompton'. Between the year ended March 2010 and the year ended March 2020 we sold, to the nearest ten thousand, 450,000 bicycles. We also sell spare parts and 'ancillary goods' – bags, clothes and so on – but if we ignore those, the total revenue over the same period represents almost exactly £700 per bike, at the point of leaving the factory.

My first experience of being on a billboard – on the side of Coutts Bank HQ in London, promoting entrepreneurship in 2012.

We can carry out the same exercise for the costs as for the revenues. In the audited accounts, the expenses are divided into two categories. The first line, 'cost of goods sold', includes the raw materials, parts and direct labour costs which go into making the bike (and into the other goods we sell: a caveat that will be dropped from now on). On this basis, the cost of making the 'decade-average Brompton' was just under £410.

But of course, you can't have a company of any meaningful size in which absolutely every penny spent is directly attributable to a specific unit of output; it costs money to organise the overall production and distribution, to have a marketing department and to do research and development, not to mention the directors' own remuneration. General administrative costs, again averaged per bicycle over the same period, were just over £230. This distinction is almost, but not quite, equivalent to the important distinction between fixed and variable costs – it's not really true to say that the cost of goods sold goes up by the same amount with every incremental bike, and nor is it true to say that the administrative costs stay the same, but it's a reasonable approximation and that's as much as you're ever going to get from a set of accounts.

So it would be possible to say that whatever the average Brompton owner of the last decade paid for their bike (a number which will vary widely depending on value added tax, import duties, transport costs and the margins earned by retailers and the distribution chain), our share was about £700. Out of this, £410 represents the amount of 'stuff' that you bought and £230 was your share of the costs of research, marketing and the other activities necessary to keep the company going. The profit made by Brompton from this abstract 'average bike' was £60, or a bit less than 10 per cent, before tax. Checking back, 450,000 bikes multiplied by £60 is £27 million, which is close to the aggregate profits we actually made (there's a small difference because we've usually had cash in the bank, on which we earned a tiny rate of interest).

Is this a good description of the economics of Brompton Bicycle Limited? Well, yes and no. We began the ten-year period over which those averages were calculated by producing 20,000 bikes a year and ended it at just below 60,000. The decade also spans the introduction of lean manufacturing and the production line system, the move to a bigger factory, a significant change in export markets, Brexit, electric bikes and a lot of other significant changes. That 'average' revenue per bike of £700 was just over £500 at the start of the period and nearly £1,000 by the end – we brought out the titanium model, the electric Brompton and several other higher value-added products, and made big changes to the distribution system which meant that we capture a lot more of the supply chain revenue. The 'decade average Brompton' is a completely fictional construct. It never existed in any actual year, even in the accounts, which are themselves a major oversimplification of the business reality.

On the other hand, as a finger exercise, it's not necessarily all that bad. The orders of magnitude on the numbers are in the right ballpark and the general structure of the calculations above tells you that it costs £X to make a bike, we take in £Y for every bike sold, and the difference between the two is of the order of 10 per cent. In engineering terms, you'd call that 'within tolerance', because it's a wide enough margin to allow a bit of space between profit and loss and to provide for investment. So although it's not really a representation of reality, the calculation captures an essential truth about Brompton, which is that financially the company is under control.

And if Warren Buffett asked you to make an elevator pitch about Brompton, your bullet points would be just that; something about how the margins worked, something about how much the company had grown and something about the investments being made for the future. Pointing at numbers in accounts and asking 'why' is a skill in its own right, and plenty of the best investors don't need to do much more than that.

I got a lot of experience in responding to people pointing at numbers during the management buyout process in 2008. The proposition was fairly straightforward if not exactly simple to execute. Andrew was going to sell down his stake so that he was no longer the majority owner. I was going to buy about half of Andrew's shares. Some of the shares Andrew sold were going to be bought back by the company, which temporarily took on some debt (it was repaid quickly) in order to neutralise the tax consequences of doing so – and, as it happened, at the time we needed to make some necessary additions to the factory in Brentford. All in, this meant that the company needed to find £2.5 million, of which I was personally going to contribute £250,000.

I did not have this money. I could never have saved a six-figure sum on the salary I had been getting, and I couldn't exactly sell my stock options to buy the shares. My only way of raising £250,000 to invest came from the fact that when I was a student I had tried to reduce my own housing costs and had been one of the first-ever customers of Barclays Bank's buy-to-let mortgage. With the encouragement of the bank managers, I had kept speculating during my time at university, using the price appreciation of one house to take out another loan and buy another. This was a new idea in the 1990s, and the next decade's worth of property boom meant that, on a generous valuation, my property portfolio had just about enough equity in it to fund my stake in Brompton. It was quite a difficult decision to take for my family finances, particularly as the property market turned out to be horribly overvalued and fell sharply soon after I had taken the loan out. I rationalised it by telling myself I was increasing control over my life rather than decreasing it – if I was ever not in a position to make the payments on the debt, it would be because something had happened to Brompton, and that would be a disaster anyway.

The rest of the money had to come from the investing public. This is the point at which people usually call in the bankers. But I didn't want to do that – it would have been expensive, and I wanted to take the opportunity to bring in investors who would help and support me and the business, rather than the top few names in the contact book of someone I'd never even met. It was a surprisingly difficult and irritating process, not through anyone's fault but just because of the mathematics of the thing. When you're trying to enlist private investors, 20 per cent is a very good success rate, which means that you need to meet at least five investors for every one you end up getting money from. And each investor will need to go back and forth with a series of questions, so the five prospects turn into twenty-five meetings, many of them involving travelling and most of them happening outside normal business hours. And you can't control the schedule. Everyone naturally wants to make their decision at the last possible moment, which means that you can't be certain how much money you've raised. First you think you've got more than you wanted, so you have to scale everyone back, and then some people pull out and you're left scrambling around to cover the shortfall. This was how I ended up in hospital. There were a lot of friends and family in the eventual list of investors, but the deal got over the line in the end.

Part of my pitch to the shareholders for the buyout was that we were going to bring in some new executives as well as expanding the non-executive board. I was aware of my own weaknesses, real and perceived, and wanted to address them. I'm a bit disorganised, constitutionally optimistic and I say yes far too much, so I wanted a CFO who would be an Eeyore – a real stickler, with a few grey hairs and a corporate approach. Unfortunately, our first full-time finance director went too far in that direction. He didn't have much experience of a fast-growing company. His methods would have been great for an environment where things happened predictably, but at Brompton he seemed to fall behind; the accounts for March weren't being completed until halfway through April. We started to lose sight of the true financial picture – although we thought we were making a profit for the year, we were actually perilously close to a loss. To his credit, the accountant quickly realised he was the wrong peg in the wrong hole and left, and the problems had not got so bad as to be catastrophic.

In itself, that was a lesson – one which hadn't been on the syllabus during the buyout masterclass – in the importance of understanding the business first and the numbers second. What's more, disagreements about the underlying business will turn into disagreements about the meaning of the numbers. When we reorganised the factory, Andrew was the founder of Brompton and I was the CEO, and we still disagreed over the question of variable costs. For nearly all of the components that go into a bike, we can either make them ourselves or get them from external suppliers. And we saw in our finger-exercise calculation above that the cost of making anything has two components – the direct cost of an individual piece, and a share of the overhead cost of an organisation that's capable of producing it.

A City financier and old university friend of Andrew Ritchie, Tim Guinness was chairman of Brompton from 2000 to 2019 and instrumental in its success.

When we make a component ourselves, the direct cost goes into the cost of goods sold, and the overhead cost goes into the overhead, as things should be. But every component that we buy rather than make is one less thing to worry about in terms of running the factory. The cost of insuring machines, heating and lighting a factory floor and all the rest of it shows up as part of the direct cost of a part if it's bought in, but as a fixed-cost overhead if we make it ourselves. As we have seen, a lot of the time there are good reasons to keep things in-house, but if none of those considerations apply and we outsource the manufacturing of a part, I can clear the bottleneck and the supply of bikes increases. This is a good thing, because it helps us produce more. But it has funny consequences for the accounts. When we buy something from an external supplier, and assuming we're buying it at a price which is consistent with them maintaining quality control and staying in business, part of their overhead cost will go into our direct cost. The balance between labour costs and materials costs will also have changed.

This is an elementary point, but it can lead to the most amazing arguments. Suppliers don't, and probably can't, provide you with a breakdown of their price list into direct costs, overhead and profit. Which means that, as happened at Brompton, if the company founder is unhappy with your factory reorganisation, they can start doing their own exercises to demonstrate that it hasn't made anything more efficient.

It's frightful to remember how much time was spent and how many board meetings were more rancorous than they needed to be in a war of duelling spreadsheets as we attempted to reconcile different estimates of the 'true' unit cost if proper adjustments were made for purchased components, wage inflation and all the other things which could be adjusted for. And it was all time wasted, because the financial numbers are just a representation, at too low resolution to convey the information that we were pretending to find in them.

There is actually no such thing as 'unit cost' in the real world. Nobody would set up a factory to make just one bike, and the 'marginal' cost of adding one more bike to a day's production tells you equally little about anything relevant to planning growth over a realistic period. It's a legitimate subject to talk about with investors and at board meetings – if overhead costs grow out of proportion to the business, they need to be challenged. But if discussion is to be conducted at any level more sophisticated than 'costs up bad, costs down good', it has to start from an existing understanding of the actual business and its processes. You can't go the other way and start reading the numbers in the hope that they'll tell you what's going on.

This is why the best way to think about these things is to recognise that money is about control, and financial debates are nearly always really management arguments being carried out at one remove. The disagreement with Andrew wasn't about direct costs and overheads, it was about reorganising the factory and outsourcing components. I believed that doing this was improving Brompton's flexibility and ability to grow – an improvement in the company's control environment. Andrew thought that it was making the bike more expensive to make, which would end up with us having to raise prices and losing control over our strategy. We couldn't convince each other about the actual business decision, so we couldn't convince each other about its representation in financial numbers.

And in all honesty the underlying disagreement wasn't necessarily all about the factory. When a relationship between two people has started out on one basis, it's very difficult to change the way they think and feel about each other even if the circumstances change. Although both Andrew and Tim had supported my promotion, in Tim's eyes, I would always be the young enthusiastic gofer who he'd met on a bus, and for Andrew I'd always be the sharp-eyed but wayward factory manager who came in and started spending money on sets of scales. Now I was the managing director; and since nobody outside the UK seemed to understand that this title meant I was in charge, I soon had to start calling myself the CEO. Even if everyone including Andrew had intellectually accepted that I had been promoted and was now going to be (albeit heavily debt-financed) a significant shareholder, that's not the same thing as accepting it emotionally. The financial restructuring which was executed as part of the buyout had brought Andrew money, and taken away control, and there has hardly ever been a company founder who has made that deal without a degree of remorse.

8 SPACE AND HOW TO GET IT

Brompton bicycles have always been made in west London. After starting out in small workshops and railway arches, a factory was set up in Chiswick. When production expanded, the company moved to Brentford, eventually needing to split operations across two sites. Currently the bicycles are made in Greenford, in premises covering 100,000 square feet.

We had to stop the line. It was getting dangerous.

Rain had been falling heavily in London for two weeks, so we had to bring deliveries into the factory rather than storing them out in the yard. We'd done our best to get everything unloaded and put away, but then two 40-foot containers arrived full of pallets of parts which had to be brought inside as well. This was in 2021, during the coronavirus pandemic, when demand for folding bikes had gone through the roof and we were making 500 bikes a day. But we had also introduced a new resource planning system, and it wasn't yet perfectly tuned. One consequence was that manufacturing had got out of sync with the sales and credit process, and we found ourselves making bikes for customers who had been put on hold because they had placed a bigger order than their current credit limit allowed. There were something like 600 bikes ready to go, but they couldn't be shipped. And there was nowhere to store them, because all the spare space on the factory floor was full of the pallets unloaded from those two containers.

Looking down from the mezzanine over the factory below, you could see that chaos was massing its forces. The pickers were having to steer carts around piles of components and practically clamber over piles of frames. Stress levels were increasing, tempers were fraying and the place was filling up with that storm's-a'coming feeling that something bad might happen. Paul Williams, our operations director, made the sensible decision to stop and regroup rather than to keep pressing on. We might have made it work, but sometimes it's better to be careful, if only to send a signal that we care about people more than maintaining production at all costs.

Having got through that day, Paul then came into the office the next morning to find that one of our delivery companies had got fed up with our Brexit paperwork. We had been arguing with them about a few delays and complications, and they had a tantrum and started refusing to ship bikes to Europe for us. There's never a dull moment in manufacturing.

Toto Ramkisoon, our No. 1 inspector, in the stores at the factory in 1998.

It was a strange reminder of the past for me. The state of the floor when Paul decided to stop the line was probably considerably tidier and more orderly than a normal day at the previous factory in Brentford when I joined Brompton in 2002. The lack of order there was one of the key reasons for hiring me; the company at that time had so much inventory that bikes were being made with some parts that had been sitting around for an entire year. This inventory was squirrelled away in odd corners around the edges of the workshop and was only slightly better organised than a teenager's bedroom. The most important components were usually to hand, but anything a little out of the ordinary could be hidden under a pile of jigs, wedged next to past prototypes and experiments.

Making 10,000 bikes is fundamentally different from making one bike 10,000 times. It's a curious and peculiarly modern phenomenon; when you scale up production, things start to be determined by overall properties of the system rather than by specific tasks, and the speed of production becomes a function of the overall design of the workflow rather than individual workers' speed of movement. All the little bits of friction and entropy start to add up. For example, you have probably never given any thought to how much time it takes to fold up a letter, put it in an envelope, seal it and attach a stamp. If you ever decide that you want to post a letter to every MP in the House of Commons, however, then the thirty-second task you haven't considered turns into more than five hours of work, just stuffing envelopes.

The problem is potentially worse than that, though. Rather than adding up, process inefficiencies often start to *multiply* up, as each delay bumps into another one, and someone has to wait a bit longer than they otherwise would have done. When you're in that sort of situation, the most important thing to realise is that the solution isn't simply to hurry up or work harder; putting more pressure on the system is just more likely to break it. What's needed is to step back and reorganise, in order to release some of that pressure and improve the overall effectiveness of the system.

What that means in turn is that as a factory grows, its overhead requirement also grows; more of the staff and resources have to be dedicated to organisation and maintaining the capability of the system, rather than directly to production. That's a slightly abstract way of thinking about things, but in a manufacturing context it has a direct and very obvious way of presenting itself. Because when you're managing a factory, rather than any other kind of business, your need for 'resources' immediately translates into shortage of physical space. When you see people crowded together and bumping into each other, paradoxically that's when you need to start thinking about devoting more space to activities that aren't directly productive.

I had a series of arguments about this in my early days at Brentford. The factory there had a mezzanine level which was mostly used to store things that weren't really needed, and I started to clear it out. I insisted that the freed-up space was going to be used as offices for non-production workers whose effort would go into organising things – and into making sure that mezzanines didn't fill up with work in progress, abandoned prototypes and general crap. At a stroke, this was taking Brompton out of its comfort zone. Adding people who weren't directly making bikes seemed like a luxury, and any small business will always be wary of adding to the monthly payroll cost without an associated revenue stream. But it had to be done; in fact, given the state of things, it almost instantly paid for itself. Anything which got the 'stock turn' (the average time that raw materials and parts spent hanging around before being turned into a bike) lower was immediately improving the cash flow of the company.

Getting to this point wasn't easy. But running out of space tends to shake you out of complacency. If you're trying to grow, then you're adding workers, and you quickly have to face the fact that every worker needs so many square feet to work in. In 2008, only two years after clearing out the mezzanine and reclaiming as much floor space as possible, we realised that we needed to build another mezzanine. This was a huge investment for us – more than £100,000 – at a time when the company profit was only slightly more than half a million. We tried to make it feel a bit special by commissioning a graffiti artist to get on a cherry-picker and create a mural showing the company history up to that date.

Even though the artist unexpectedly got vertigo, meaning that the mural didn't go quite as far up the walls as we had hoped, it was well worth the money (if memory serves, we gave him £500 and a bike). How much people's mood is affected by their work space is greatly underestimated. If somewhere is bright and spacious and looks good, people feel they should live up to it. But if you feel like you can't bring your kids to work and you have to queue for fifteen minutes to go to the toilet, it drags you down.

In Brentford, though, the space was gradually strangling us. The new mezzanine wasn't enough. We had dozens of ideas that we couldn't implement for lack of room. We had enough demand to run two production lines, but there was only room for one, so we ended up having to put on a second shift, paying time-and-a-third for weekend working. We installed some mobile offices in the yard, moving the canteen into one of them, but I was increasingly aware that this sort of thing wasn't going to help – trying to solve an overall system-level problem by throwing more resources at it.

We needed to move to a bigger factory and to completely re-arrange the workflow, starting once again from first principles and respecting the changing scale of our output. By now it was 2012, heading into 2013. We were making just over 40,000 bikes a year, and we needed to think about how things would be organised at some point in the future when we were doing multiples of that. This was going to be difficult, and risky. When small and medium-sized businesses go bust, very often the failure has been immediately preceded by a site move. It's easy to see why. If you're moving, it's usually because you're growing, so you will want to move to a space where you've got room to

HRH the Duke of Edinburgh visiting the Brentford factory in 2010. He later officially opened the Greenford factory in 2016.

grow some more. For a while, therefore, you are trading a space that's too small for one that's too big; the overhead and rent cost becomes a bigger obligation every month compared with the size of your business. Add in the cost of the move itself, and the inevitable disruption and diversion of management attention, and you can see why this turns into a period of heightened vulnerability, in which any ill-timed piece of bad luck on the revenue side can trigger a crisis.

It was also going to be difficult finding a bigger factory because not many sites are right for a business like Brompton, particularly in London. Most so-called 'industrial' buildings are just sheds. They're suitable for a logistics or warehousing business without much adjustment, but we need to bring in oxygen and acetylene supplies, and to have extraction; we also need offices next to the factory floor. What this means for a manufacturing business is that you end up

In the new brazing layout, stations are designed ergonomically for each specific task.

spending a lot of money on fitting out a building that you don't own, and which nobody is going to pay you back for at the end of the lease – on the contrary, you will usually be made to pay to remove what you've put in.

So ideally we wanted somewhere that we could own. That would mean a lot of money; we looked at sites outside London to reduce the cost, but if we had moved too far from west London we might not have been able to persuade our staff to follow us. This would have been deal-breakingly disruptive, because there were some rigid facts built into the business model; if we lost half our brazers because they didn't want to move house, it would take years just to get back to the old level of output, let alone expand. Just as importantly for the company's DNA, we couldn't move to an industrial estate on the outskirts of London. We need our people to be able to ride their Bromptons to work; this was the stimulus for Andrew to design the bike in the first place, and it's the way that we drive innovation and improvement.

All these reasons made us conclude that we would have to get into the commercial property business. Brompton had one trump card when it came to London real estate – we offered good manufacturing jobs for local people. We had an idea that we could partner with a property developer, on the basis that they would find a site that was large enough for a factory and some flats. We would then approach the local authority together and apply for a change of use to permit the site to be mixed industrial and residential – effectively our contribution to the project would be rewarded by getting our factory site at a reasonably affordable price.

We went through a couple of versions of this project over a couple of years, but always seemed to fail at the last hurdle. The problem in dealing with a developer is two-fold. First, the land isn't going anywhere; and second, once something has been built, it's pretty difficult to build anything else on the same piece of land. The first point means that the developer can usually afford to wait longer than anyone else, and the second means that they're very reluctant to compromise while they think the most profitable deal might still be possible. We had a totally different agenda, since we quite urgently needed a deal, so we were vulnerable to people taking us on and then changing their terms at the last minute after everything had been agreed; conversely, if the local authority decided not to give permission for the number of flats the developer wanted, we weren't able to wait them out.

This is a general problem which managers of companies of all sorts may recognise, and it may account for some of the ill feeling between industrialists and the world of finance and real estate. When you're trying to grow a company, space is never just square metres and location – as I've said, shortage of physical space stands in for many of the problems of complexity and scale which go along with growth. Similarly, money is never just money – it's part of the same equation. For people who deal in finance and property every day, it's normal to consider time, space and cash as just other kinds of commodity, but that's not possible for a company like Brompton.

Learning this lesson was slow and frustrating; we had started thinking about it in 2011, and the search had gone on for four years. Grudgingly, and knowing it wasn't the ideal solution, I admitted that we were going to have to rent somewhere. By this point, the capacity constraints in Brentford had become absolutely unbearable, as space, scale and complexity started feeding off each other in the most toxic ways imaginable. We had been reduced to splitting between two sites, with the dispatching, spare parts and some customer services consigned to 'Unit 19', a space tucked away behind the 'Golden Mile' of the A4 highway. In principle, Unit 19 was only ten minutes away by bike, but what happened was all too predictable. My best intentions of regularly going there never worked out, the staff who were moved there started to feel marginalised and ignored, and any time something went wrong or got lost, there was

Fred Torto, our first brazing team leader, using the chain stay setting tool, Brentford, 2003.

always a low-level feeling of uncertainty about how to sort it out. It's impossible to expect people to produce their best work in those circumstances.

We narrowed the choices down to two options. One was very competitively priced and conveniently located, but it was only 65,000 square feet; enough to amalgamate our current split, but not really providing room to grow. And the other was the site in Greenford. That was 100,000 square feet, somewhat more than twice our existing space. This would be very big – it would allow us to bring in operations that we were outsourcing for reasons of space rather than desire, like wheel lacing and the paint shop. But even with those operations added, we would still be paying, in the short term at least, for a lot of space that we weren't filling. This was a tough decision – we even went as far as holding a full board meeting at the smaller location before deciding that we had to take the risk.

But there was one very powerful equaliser on our side. And that was the building itself. Within twelve months of starting up there,

A final goodbye to Brentford before heading off to Greenford in 2016.

we had already made back more than the cost of the additional rent from efficiency improvements that people came up with, and which we were able to put into practice because we finally had enough space to organise the line.

Some of the improvements were big, easy wins – we were able to put two production lines side by side and cancel the weekend shift. But it also turned out that I'd underestimated the benefit of insourcing, both from savings on the suppliers' overhead and profit margin, and from optimising processes for Brompton bicycles. We found that we were saving a few pounds per bike on paint costs, a few more on wheels – these benefits start to add up quite quickly.

Things kept getting better, because the real benefits of the new space were hard to put into any business plan – they came from new ideas and the ambition that the building inspired in all of us. The Brompton factory in Unit 1 of Greenford Park isn't exactly a beautiful building from the outside, but it is light and airy and it felt so big when we walked into it for the first time. Just like when we put in the mezzanine and had the mural painted in Brentford, people truly felt that they had to live up to the space they were in. We had been talking about reaching 50,000 bikes a year for quite some time, but I don't think anyone really believed it until they could look around them and see a factory that was obviously capable of doing that and potentially a lot more.

For example, our assembly line is computerised. An assembler arrives at their station, then logs in. The system first of all checks that the operator is trained on that station, and that nothing has changed too much since they last performed these assembly operations. The assembler turns around to pick up the frame, and the station screen tells them that they ought to be seeing a part-assembled pink and green Brompton, because the order that they are working on was created in those colours on the website, and the frame has been brazed and painted. After confirming that the picker who loaded the frames got it right, the operator scans the serial number of the frame so that the system knows who is working on it.

These screens at every station are key components of the system, but they cost us less than a mobile phone; £40 touchscreens driven by a £22 Raspberry Pi (a chip that replicates an IBM PC, initially sold for educational purposes but bought in bulk by anyone who needs cheap computer control).

Say that the job at this station is to fit the bottom bracket to the frame, using a torque wrench. The bottom bracket is a safety-critical component, so it's important to monitor that it's been tightened properly. But you can buy Bluetooth-enabled torque wrenches these days, so the system can log exactly how much torque has been applied. The system also keeps track of how long the job takes and when the next frame is scanned. This data is valuable because 'balancing' the line is never a once-and-forever task; we need to keep checking that the time spent at each station is consistent, or work in progress will start piling up somewhere.

Each individual bicycle now has an identity in the system throughout its assembly. That's useful in itself because it lets us plan our shipping and communicate with customers. The system also tells the pickers which stations need which parts and when; it's a leap forward technologically, although we still use physical cards to track things. And from time to time, we can aggregate data in order to see what patterns emerge – if there's a particular station that is always slowing things down, for example, or if more people need to be trained on one stage of the process in order to stop bottlenecks happening.

Manufacturing control systems like this are usually seen in the auto industry; it's quite unusual to have one in a medium-sized bicycle factory. Installing them is very expensive, requiring a big software licence and usually several dozen consultants to customise everything. We have in fact spent seven figures ourselves on external resource planning software, which we use to deal with inventory and component ordering, but we didn't need to spend another seven figures on customising the vendor's system to integrate it into our factory – it just performs its function and then hands over its output to the Brompton control system. And our system was all built in-house, on the initiative of Brompton employees who came up with the ideas.

Martin Best finishing a main frame in Brentford, 2014.

And that's the final lesson of the whole episode for me; there are ways in which the feedback between space, scale and complexity can turn into a positive spiral rather than a vicious one. When you're constrained, everything turns into a problem; you want to produce more, but that gets more complicated, and needs more space, and that creates problems which take up your time and energy trying to solve, making you feel even more constrained as the output stagnates. If you can execute a step-change, though, and buy yourself enough space – both literal and mental – to consider the system as a whole, then you get a chance to create an environment in which growth brings you solutions rather than problems. If people aren't being ground down by the pressure of an impossible situation, they can improve things. It's a similar phenomenon to what we saw earlier, when we were thinking about organising processes to build bikes. You don't solve problems by addressing specific problems; you solve them by looking at the whole system in which they're embedded, and coming up with a way to let the system help itself.

9 DEALING WITH DIRECTORS

> *Will's first title at Brompton was 'new projects manager'. Since joining the company he has been engineering director and managing director and is now CEO.*

Chris Boardman, the cycling champion, has a way of explaining how he makes plans while executing them. When he's doing a time trial stage, he's on his own. He doesn't know how fast his rivals are going or what he needs to do; all he's aware of is the time and distance covered and how he feels at his current pace. So he asks himself, 'Can I keep this up for the rest of the stage?' If the answer is 'Definitely no', then that's bad news and requires an immediate change of plan. But if the answer is 'Definitely yes', that's also bad news, because nobody wins a stage of the Tour de France by staying in their comfort zone. So he needs to be in a place where the answer to that question is 'Maybe'.

That sort of thinking was on my mind all through the process described in the last chapter. The factory move was a story that made sense in industrial and business terms. But there was another story going on at the same time, which people didn't see so much of because there was no reason for them to. All this time, I've been talking about getting things done and managing increasing complexity by building systems to deliver results. And one of the most important examples of such a structure is the board of directors of a company. Lots of people have calibrated a CNC machine, but maybe fewer have been the chief executive officer of a company and needed the support of their board for a major decision. So it's worth going into this a little.

As a manager, building and maintaining relationships with directors and shareholders can take you quite a long way out of your comfort zone. Not only is it time-consuming, it's also a very different skill set from managing the daily operations of a factory or any other business. It's tempting to tell yourself that you're just going to get on with the job in hand and that you don't play politics. Even more than that, the business of building a consensus for a board decision can often be really unpleasant, tough work. There's a lot of eating humble pie, a lot of allowing other people to take credit for ideas. And a lot of what in many ways is the toughest job in business, something that many people never manage to achieve at all despite having every other ability – sitting down, being quiet and listening to other people's opinions, without answering back.

But if you don't pay attention to your board and shareholders, you lose a lot. For one thing, your knowledge will contract. Most employees won't directly say no to the CEO. Even customers and suppliers, who don't literally work for you, will tend to dress things up nicely and present information in such a way as to make it easy to ignore the nasty bits. But non-executive directors and shareholders can tell you to your face when they don't agree with you. And you need to pay attention to them, because they can stop things from happening. A board of directors will not generally micromanage and second-guess everyday operational decisions, so if you have a communication problem or a difference of opinion with them, it won't become immediately apparent in a series of disagreements on the shop floor. Instead, all the problems get stored up and released onto you in one load. If you haven't prepared the ground, explained your case, made allies and understood people's concerns, then you'll find yourself with a big, company-defining strategic decision to make but being unable to make it.

After the management buyout in 2008, my new relationship with the board didn't start off particularly well. It was awkward, as I've described, for people who had known me from the day I joined Brompton to accept that the young engineer and dogsbody they had hired was now in charge. I knew that if I were to have a chance of succeeding, I had to insist as part of the deal that Andrew would keep his board seat but step back from operational control. This was agreed on all sides – it was what he wanted anyway to reduce his own stress and workload, and it was the only way that I thought the company could function. He remained in the office, working on a sophisticated pricing algorithm and on organising and codifying the design philosophy, but away from all the details.

Unfortunately, this meant that he would find out about things late in the day, in briefing packs for board meetings or – even worse – during the meetings themselves. At this point, all the decisions which I had taken would be revisited. All the detail and operational minutiae of the previous month would come back in one lump, usually when it was too late to do anything about them except argue. I doubt if I handled this as well as I could have done, but board meetings became an absolute trial for me; it was a sort of Prime Minister's Question Time in which I played the part of an embattled PM in the middle of an unpopular war. Everyone fired questions at me until they found a gap in my knowledge, and then spent the next fifteen minutes excoriating me for it.

Over time, I worked to change the atmosphere and composition. It had been all right to keep things fairly informal when there were relatively few investors and most of them knew each other, but by the time I took over in 2008, Brompton was becoming a medium-sized company, with a few million pounds of other people's money invested in it; it felt like we needed to take things seriously. We started to have formal agendas and take minutes. We were also able to bring in some excellent non-executive directors. John Putt was an automobile and motorsport executive who had simply written to us saying that he had tried to order a bike and been quoted a sixteen-week lead time, and that we 'needed his help'. We had a vacancy on the board, and I couldn't really argue with his point about the lead time, so he came in. Things really changed, though, when we added Jo Staveley, the retail executive who had brought Kath Kidston and LK Bennett to the high street. The board members were all better behaved when Jo was there, partly because she was a woman, but mostly because she was considerably more successful than any of us.

Working on the board composition like this was weirdly reminscent of fine-tuning the automatic brazing machine, getting the mixture just right, attending to the temperature and watching for anything that might throw the atmosphere out. In many ways, the problems and strategies that you have to come up with in order to function as CEO reprise a lot of the principles and lessons learned from manufacturing itself. The perspective shift you have to make is that when you're dealing with materials and machines, all of your

The original shareholder register of 1979, made up broadly of Andrew Ritchie's friends, most of whom remain shareholders to this day.

No. 1. 200

BROMPTON BICYCLE

LIMITED

Incorporated under the Companies Act, 1948

CAPITAL - - - - - - £100,000

Divided into 100,000 £1 ORDINARY SHARES

This is to Certify that J. D. ILIFFE of 75 STANHOPE MEWS EAST, LONDON, S.W.7.

is the Registered Proprietor of TWO HUNDRED Shares of £ [redacted] each, in the above Company, numbered from [redacted] to [redacted] inclusive, subject to the Memorandum and Articles of Association of the Co[mpany] and that upon each of such Shares the sum of £ [redacted] has been paid.

Given under the Common Seal of the above Company, this 23rd day of June 1976

A. Whitley } Directors

George Brocklebank Secretary

NOTE.—No transfer of any of the above-mentioned Shares can be registered until this Certificate has been deposited at the Company's Registered Office.

Joe Iliffe's 'number one' share certificate.

problems are to do with uncertainty and tolerance and you adjust your processes according to your controls. At the organisational level, your control itself is one of the uncertain factors and you have to work to ensure you have the resources to exercise it. This is particularly obvious when, as we have seen, questions of control show up in the form of questions of money. In other contexts, though, there's no clear measure of your ability to make decisions about the future. As a CEO dealing with the board, I was left calibrating things by personal relationships and logical argument. I guess if you wanted to give this approach a name you'd call it influence, and it works within a structure just like any of the other systems we've looked at.

The shareholders own the company in proportion to the number of shares owned. In principle, the company's charter is that of a constitutional democracy, where decisions are at least potentially made on the basis of one share, one vote. In practice, this rarely happens. Either the shareholders are a small group of people, as at Brompton in the early days, in which case things have to be done mainly by consensus because the shareholders, directors and management are largely the same group of people, and they have to get on with one another. Or when the company is even a little bit larger, it has too many things going on for the shareholders to be able to exercise any meaningful control, given that they generally have lots of investments and only limited time and attention.

Because of this, it's intrinsic to the concept of a company that it will have a board of directors, to whom most of the decision-making power is delegated. Again, in very small companies there will be a significant overlap between the directors and the management, but as the firm grows, it's expected that 'non-executive' directors are brought in. These are people who serve on the board, but don't have any close connection to the operational management of the company – they're meant to exercise oversight and to slightly mitigate the huge information advantage that the management have, which would otherwise allow them to do more or less whatever they liked without any governance from the shareholders. Since you have to have non-execs, it makes sense to get good ones, who can add a bit more value than that; people with relevant experience in manufacturing and retailing, or financial specialists who can get behind the numbers. It was a significant measure of Tim Guinness's ambition for the company that he started bringing in non-executive directors right back in 2000, at a much earlier stage than many smaller companies would have bothered to do.

There's a hierarchy of decision making. Most things that happen in the company from day to day are handled by the managers, with the CEO at the top of the pyramid. For big decisions, I need to get approval from the board. I can decide for myself what constitutes a big decision, but board members see a certain level of reporting and information and they can put whatever they want on the agenda. And at the level above that, there are matters that have to be approved by a vote of the shareholders. These are basically anything which affects their ownership stake: dividends, capital issues and mergers and acquisitions. For the annual dividend, it's usually a formality – the proposed dividend is based on what we can afford, and it's very unusual for anyone to question it. But the shareholder vote is important for capital increases and changes of control, and understanding the interests involved is crucial if you want to win or keep control of where the company is going.

Winning the highly competitive and inspirational UK HSBC Connections competition, 2013.

The factory move was a big decision. The board was split. I had the advantage that Andrew agreed with the move, as he was very much in favour of a site that would allow us to do our own painting and wheel lacing. Tim, however, came from a financial background and was well aware of how many businesses moved to a larger factory and then folded. From his point of view, the Greenford site would commit Brompton to renting much more space than we would be immediately able to use, as well as the expense of fitting out.

And however constructive and friendly the non-executives were, they had a duty to the shareholders to hold me to account. The company was not going to take on debt, but the dividend would be frozen at a relatively modest level for a few years, and the expense involved would significantly reduce profits. Sales revenues had been growing strongly for years, but there was no guarantee this would go on for ever. If revenues were to falter, the rental payments on the new factory would start to feel a lot like debt, draining the company's cash and reducing its operational freedom. This concern was well-founded. The reason why sales revenues had been growing had little to do with the number of bikes; they were growing because we were buying out many of our overseas distributors, and so keeping a larger proportion of the sale price. The number of bicycles shipped had stagnated at around 44,000 units a year. This was partly, in my view, because Brompton had been looking inward at itself. We had been thinking about the organisation of the factory, the management structure and the commercial network – all of them important things, but the pace of innovation had slowed for the actual bike. That's always a difficult balance to strike, and we needed to rebalance things quickly.

It was all, obviously, going to depend on the business plan. Broadly speaking, if I was too conservative in my projections, then they wouldn't justify spending the money, and the investment wouldn't be approved. At the other end of the scale, everyone knows that if you don't care about the future and promise the sky, you can justify more or less anything you want. In the space between those two alternatives – that's where the interesting decisions have to be made. I decided I was going to follow Chris Boardman and come up with a plan that had me thinking 'maybe'.

As Brompton became a British manufacturing success story, politicians often visited the factory to learn something about export and manufacturing – and of course for a photo opportunity!

It wasn't exaggerated or manipulated – I was perfectly content to look everyone in the eye and tell them it was what we planned to do. But it wasn't exactly in the comfort zone either. The sales and management people who would have to deliver on these plans had gone back and forth with me a couple of times to ensure that we were making our best case, and I think I'd admit, with hindsight, that it was a document produced to achieve a result. And given the way that things work, it was quite likely that, in the initial stages at least, my projections were more likely to understate than overstate the challenges, and that consequently the period immediately following the decision was likely to be followed by some rough board meetings.

So I also suggested that, since the board didn't all support the plan, there ought to be some method for people who were against it to take their money out of the company. If Brompton was quoted on the stock exchange, this would be simple; anyone who didn't like it could have sold their shares. The share price might have taken a temporary knock, but it wouldn't have affected my ability to keep running the business. For a private company, though, this was considerably more difficult to manage. Once again I had to go and find new investors willing to buy in, and at a sufficiently attractive price to persuade the incumbents to sell rather than staying on the shareholders' register and potentially voting against the move.

Having given myself pneumonia the last time I tried to find investors for Brompton, I wasn't exactly joyful about returning to the endless round of meetings. But there was a purpose in adding this option to sell up and leave; as always, I was trying to treat the problem respectfully by looking at the value generated and reasoning backwards through cause and effect. I knew that some people on the board benefited from favourable tax treatment relating to their long-term shareholding and the fact that they had backed Brompton from the earliest days. If they could be persuaded to sell to the new investors, then the amount of money they would realise could potentially be quite substantial. But the new investors would only come in on the basis of the new plan, which included not only the move to Greenford but also a big financial investment in taking development of an electric Brompton beyond the prototype stage. So anyone on the board who wanted to take advantage of the exit terms now had a strong incentive to support the plan.

This was quite a political step forward for me. To begin with, it had looked like the plan was simply a matter of me on one side and the financial investors on the other. Then it won the support of Andrew, because he believed in the factory move. And now some people who might have been concerned about the financial risks started to think that supporting the plan was in their best interest in order to get a good exit price. The cautious faction began to crumble, and the factory move started getting that air of inevitability which always brings people on side.

And thankfully, by 2015 it was a lot less painful finding investors than in the days of my first hotel presentations. The company was better known, and so was I; there were nearly another 200,000 Brompton owners in the world; and the global financial crisis had receded into the past. A year or so earlier, I had attended a government reception for a 'Best of British' trade promotion event, the kind of thing that's always going on in some corner of the civil service. At these events you can often tell who is more important than anyone else by the knot of people around them for the whole evening. On this occasion, the cluster was around Josh Berger, the president and managing director of Warner Brothers UK. I didn't join the huddle at the time, I just mingled and enjoyed the smoked salmon and English sparkling wine. But behind him in the cloakroom queue afterwards, I couldn't miss it when he asked for, and received, his Brompton.

I had to catch him up, obviously, and started a conversation at the next set of traffic lights with a nonchalant 'nice bike'. As we started to chat, Josh revealed that he was such a fan that he had written to Andrew to say what a great invention it was. (I checked back in the office, and he had.) When we were looking for new shareholders

Josh Berger (CEO of Warner Bros UK) in 2014 with his custom-made 'Academy Red' Brompton, which he would ride up the red carpet for film premieres.

a year later, he was an obvious candidate to approach, along with Dan Cobley, at that time the UK managing director of Google, the venture capitalist Luke Johnson and various others. Josh ended up becoming a great friend of the company as well as of me personally; to say thank you for the help he gave us with introductions in the American market, we made him a custom Brompton painted in 'Academy Red', the shade used for the Oscars carpet.

In a different life, I think I might have made an investment banker. I like meeting people and networking, and I don't forget contacts that I make as soon as the door shuts. Dealing with shareholders and directors is one of the things that people make more complicated than it needs to be. It's fundamentally about building trust, and that comes from two things; being honest, and remembering to communicate. If anyone is surprised by something they find out at a board meeting, that's a failure, either on their part or mine. It means either that they haven't done the work to be prepared for the meeting, or that I haven't made enough of an effort to make sure they understand what I'm planning.

The thing about this sort of relationship is that by the time you need to carry the board with you, it's too late to start thinking how to do it. You can't start planting seeds at harvest time. There have been a few occasions – only a very few, and none of them pleasant – when I have had to work to remove a director from the board at Brompton. When it has happened, it has come as a total surprise to the person I was getting rid of, because they either didn't believe I would do it, or more often believed that I couldn't because the other directors and strategic shareholders wouldn't support me. I can certainly see how people get addicted to boardroom politics. It's boring for the most part, and often tiring and awkward, but there is an undeniable thrill in seeing a plan come together and winning a vote.

Winning the the vote on the factory move and successfully executing the share trades didn't feel like such a victory a few months later on, though. The Greenford plan was probably about as successful as I could have hoped, and certainly more so than even the optimistic projections I'd given to the board and the shareholders. But of course, with anything like this the costs are taken up front and the benefits accrue over time. We had monthly board meetings, and for the first three or four of them after the big vote, the numbers did not look great. Sales were still flat, the overhead expense had gone up and the variable costs were a bit higher too, as we got used to the new factory and bedded in the processes.

I was soon getting used to hearing from board members that Brompton was the worst investment they had ever made, that the company was broken beyond repair and that my business plan was a pack of lies. Personally, I think this kind of rhetorical flourish doesn't add anything to the point someone's trying to make, but I had to grin and bear it. (I began to push back in the end because I realised that I was no longer bringing managers to board meetings because I didn't want to be responsible for what happened to them there.) Even that may have been a useful learning experience, though.

The faces change, but the structures stay the same. Board meetings are *meant* to be tough for the CEO. That's their function within the system; you can make them polite and friendly, or confrontational and shouty, but the conflict is intrinsic to the system that's been built. It's a system that's designed around relationships and human beings rather than machines and materials, but just like the manufacturing process it's all about control, uncertainty and tolerance and so the same principles are true. A functioning system is one which generates solutions and information by the way it's set up; a non-functioning system tries to deal with problems one by one by expending individual effort on them.

10 THE WHOLE PERSON

Brompton in 2022 has over 800 employees. More than a quarter of them joined the company in the previous two years.

I had just got off the phone to one of our suppliers, telling them there was a fault in one of their machines. A steel press in their factory had got a little bit of grit in it, which was occasionally causing a small blemish in some of the parts that they supplied to Brompton. This blemish was more or less invisible to the naked eye, but it showed up when the completed frame was painted. That was when I realised we needed to create a job for a full-time computer programmer, and that the job needed to be given to Kane Tracey.

This was a slightly strange move; it was around the start of 2012, when the company had fewer than 120 employees in total. And although Kane had previously worked in IT, he wasn't a software developer at the time; he was our chief paint inspector. But it may actually have been the best decision Brompton made that year. I wanted to create the new job because Kane had, on his own initiative, set up a paint inspection system that was now allowing us to diagnose faults in other people's machines. If someone does something like that, it makes sense to see what else they might come up with.

If this had started in my office, I would have needed to justify hiring a programmer; it's the sort of decision that might even have been discussed with the board. To explain why we needed a programmer, I would have had to come up with a list of projects which I thought we needed a programmer to do. Then whoever we hired would have felt they had to complete that list, or at least make substantial progress on it, before having any ideas of their own. The paint inspection system wouldn't have been on the list. And that would have made a substantial difference to the company today, because a surprising number of other innovations over the last ten years can trace some of their origin back to that system. It's worth going into a bit of detail here.

The paint and finish are important on a Brompton, because there's a particular way that people relate to a new bicycle. It's a considered purchase, it's got a certain image, and we encourage everyone to treat it as if it was a little bit special. That means that at that first moment of ownership, the customer wants it to look absolutely perfect, like an iPhone. Even if literally ten minutes later they're going to be riding it down streets, through puddles and over broken glass, that unboxing moment has to happen. Any little defect is going to stick in people's minds; if they happen to see something wrong under the lighting in a shop model, they will either haggle over the price or not buy at all. So we have to meet that expectation with high standards.

Unlike the assembly line, the paint shop is pretty much a discrete operation and there's always more potential for variability. Because you're dealing with viscous fluids rather than solid metal parts, paint and finish still need to be inspected; there's much less scope to design quality in, so you have to inspect defects out. There is a whole language of paint inspection. For the Brompton paint shop, the two biggest evils have always been orange peel and inclusions. An 'inclusion' is a piece of dirt or grit that hasn't been removed from the frame or which was floating in the air and which has got 'included' in the coating. And you would know 'orange peel' if you saw it, because that's what it looks like; an unevenness to the coating which can have a number of causes but which, in the powder-coating system used for Brompton frames, was typically a result

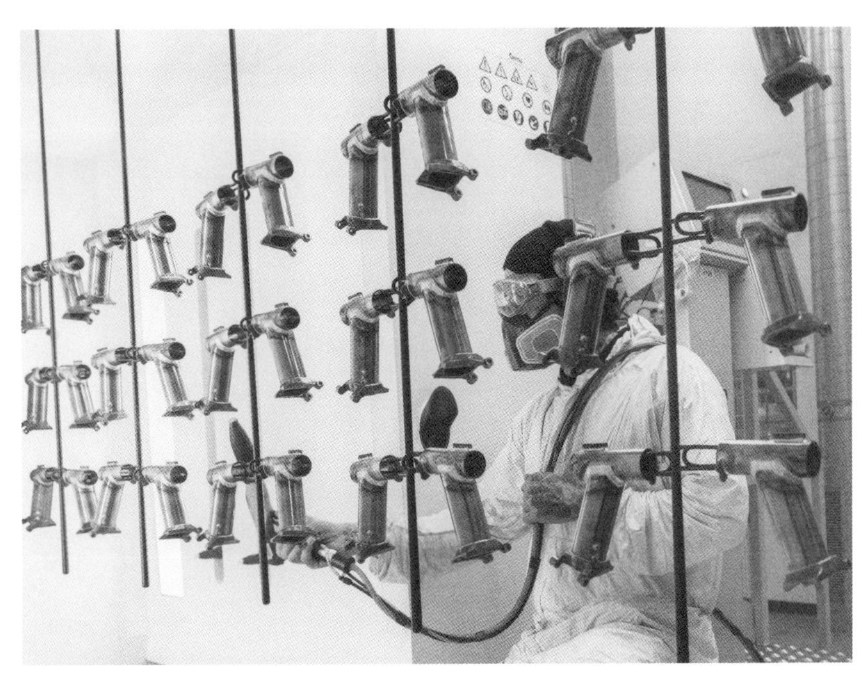

The paint plant in the Greenford factory. Previous factories had not been able to accommodate the space needed.

of surface contamination by some kind of residue that the coating couldn't stick to properly.

Some other paint-shop defects such as runs or drips will never be found on a Brompton, because they're associated with the characteristics of liquid paint. We powder-coat the frames, which involves spraying plastic powder from a paint gun which has a high voltage in the barrel. The tiny particles of plastic acquire an electric charge as they are sprayed out, so they will tend to stick, electrostatically, to an earthed metal object, like the bicycle frame they are being sprayed onto. As a bonus, because of the physics of electric fields, the charge will be stronger on convex surfaces than concave ones. This means that paint particles aren't drawn into the inside of the frame tubes. The electrostatic paint gun leaves the frame covered in an even layer of powder; it's then taken into an oven to melt the powder into a smooth coating and gradually cooled, trying to make

The complicated mechanism behind the powder-coating paint system.

sure that nothing goes near it while it's cooling that might result in an inclusion.

This requires specialised equipment and quite a lot of space, so before the move to Greenford we used to contract out our painting to a factory in Wales. Outsourcing the work inevitably made things more difficult, as handing things off and then taking them back always does. It was not always entirely clear where a defect had occurred; whether an inclusion had got into the coating during the painting (or cooling) process, or whether it had been caused by an imperfection in the frame, for example. And the fact that everything had to be shipped on a 200-mile journey made mistakes significantly more expensive, in time as well as transport cost. If, at a final inspection, someone spotted a blemish, there was nothing to do but take the whole bicycle apart, hinge and all, then send the errant part off for a difficult process of removing the coating, reapplying the corrosion protection, cleaning, coating, curing and cooling. It was then doubly frustrating if the problem turned out to be with the frame after all.

A consequence of this was that a kind of 'grade inflation' began to set in among the quality control team. The cost of a mistake was higher the later in the process it was spotted. So when things were found at final inspections, there had to be a degree of accountability back to the earlier inspectors, who got into trouble. People therefore began to inspect things defensively to avoid anything being declared their fault. The smallest imperfection, and something would be sent back for a respray.

I did the wrong thing – I tried to solve the problem. Over a period of weeks, sitting down with the quality assurers and looking at feedback from the sales channels, I decided what the paint quality standard was going to be. This was literally a process of codifying everything. We assigned every part to Class 1, 2 or 3, depending on how visible it was, and set standards for the size of inclusion that could be permitted in each area, the amount of orange peel that could be tolerated at different points on the frame, and so on.

By the time this was finished, the paint standard filled up a binder. Was it a good standard? Yes. More or less, it's the same system that we use today, and as a piece of work it was both necessary and good. Did it solve the problem? Not at all, because the problem fundamentally wasn't about paint, it was about people. Within weeks, it seemed like everyone had their own interpretation of what the standard meant, definitions were drifting and neither the excessive rejection rate nor the inspection paranoia was any different.

At this point, Kane Tracey entered the story. We had decided that what we needed was a 'paint tsar' to be a single point of knowledge about paint inspection and hopefully to be a transparent arbiter of all the questions that were being left to rumour and assumption. As it turned out, Brompton got a lot more than we could have hoped for. Kane was a good paint inspector. He was capable and diligent, but more than that, he was enthusiastic and so I got on well with him. When he came to ask if he could borrow some frame parts to try out a project for improving the paint process, I had no real expectations, but he'd made a good impression on me.

The reward was pretty extraordinary. Two or three weeks later, I was standing in front of Kane's computer, as he showed me a screen with rotating 3D models of all the Brompton frame parts. If you clicked on any part of the frame, a drop-down menu appeared, giving you the relevant section of the paint inspection standards folder, and inviting you to select the precise part of the standard which hadn't been met, the size of the fault and whether it had been caused by the handling or the supplier.

This tool turned out to be almost revolutionary. In terms of ergonomics and ease of use, it was much better than pen and paper. The important thing, though, was that the menus forced you, as an inspector, to be precise about exactly what was wrong with the frame in front of you, without ambiguity. There were no longer any borderline cases. And as the data was logged and built up, it became possible to see patterns. Rather than allowing rumours to spread

Handlebar supports stored in bespoke boxes ready for assembly into frames.

about differing standards, we could simply see how some inspectors had lower tolerance thresholds for orange-peel effects, or higher sensitivity to inclusions on certain parts, and we could start targeting training to standardise judgements.

This was exactly what we needed; the inspection process itself was generating the information needed in order to improve the inspection process. As we built up more data, the paint inspection system was able to improve other processes too, like warning us of the grit in the supplier's press or telling us that we were handling the steel tubes wrongly. We also now had those 3D models, which formed the basis of the 'configurator' on our website, an incredibly popular feature that lets people pick their options and colours to see what their bike will look like. And the standardised description files used by the configurator form the basis for the manufacturing control system which lets us track bikes through the assembly line. And that system, in turn, gives us more and more data about the process, which we can use to spot other problems and improve efficiency. The ideas have just kept flowing, and it all started because we had a paint inspector who brought his passion for computers into his day job.

It's like climbing a mountain. As an expedition leader, you can choose which mountain you want to climb, and find out what kind of climb it's going to be. Based on that, you can go out to find the people who have the right skills and experience. But once you've recruited them, you don't expect them to ask you what route they should pick, or what equipment you want them to pack; that's their role. If you're an expert on ice climbing, then once you've made the decision to tackle a hot rock peak in Nevada, you know that most of the rest of the decisions are going to be made by the other members of the team. Although it seems obvious in that context, lots of people arrive on their first day in a new job and ask the boss what they should do. Too many bosses then tell them, instead of giving the correct answer, which is something along the lines of: 'That's your job! I can tell you what we are trying to achieve, but you need to decide what you are going to do to help us achieve it. We spent ages looking for you, exactly because you have the knowledge and experience that we don't have.'

In fact, even hiring a new person in the first place is not always the best idea. Sometimes it's necessary, in order to bring in talents and experience we don't have. But it should always be a last resort, only after we've really thoroughly looked to see if someone already working for Brompton has that talent and experience. That may involve thinking a bit laterally, looking beyond the role that someone is working in at the moment. The specific job that you hire someone to do is really very unlikely to account for more than 10 per cent of their personality or ability, so if you just think about someone in terms of what you see at work, you're missing far more of their potential. If you take the trouble to look out, ask questions and be interested in the human being, you're expanding your effective talent pool by an order of magnitude.

When we realised that we needed to have a permanent organisation on the ground in the USA, for example – and when I decided that I just couldn't move my family to do it myself – we asked Katharine Horsman, one of our marketing managers, to run North America for us. She agreed, as long as she could first take a sabbatical to travel with her then boyfriend, Rich Spencer. Rich was responsible for training people on the shop floor and maintaining standards and manuals, and I suggested that he might go along with her to represent the company in New York; from running the training, he knew most parts of the bike intimately, so he would make a good operations director. Katharine eventually left for a career in marketing (she's now a consultant), but she and Richard were the first and second presidents of Brompton USA, and Richard ended up becoming one of America's top cycle retailing experts.

At one level, this might not seem a particularly controversial philosophy. After all, everyone wants to say they hire exceptional people, create an open culture in which employees feel free to express ideas, and trust their staff to innovate. (There are not many managers who will say out loud that they would prefer to recruit zombies that they don't trust, then force them to do exactly what they're told and no more.) But I hope I've made it clear that when I'm talking about what we do at Brompton, it's not hot air. Our sub-assembly team manager is Craig Knight, a young guy who joined us at the age of sixteen and immediately stood out for being unusually diligent and enthusiastic. He never seemed to want to go into the brazing school, but he had exactly that quality which Abdul and Rebecca recognise of wanting to listen to advice, and determination to do things right. So we started to ask him to do extra projects for me, away from the assembly floor. From time to time I might even have overloaded him, working on the basis that if you want something done, you give it to the busiest person.

But it wasn't long before everyone began to notice that Craig had an unusual eye for the assembly process. In the way that some people can do jigsaws really quickly, Craig notices inefficient movements,

Early days in the USA at the Las Vegas bike show 2007, with our very simple but practical stand. From left: myself, Michelle Mitchell, Katharine Horsman and Emerson Roberts.

bad ergonomics and unbalanced workloads. He says that he will sometimes get ideas while shaving or on the way to work, suddenly realising that something has been bugging him for the last few days and that he's realised how it can be made better.

When you have someone with a talent like that, it makes sense to feed it. So we started sending him on courses in lean manufacturing techniques, to give this intuition more of a structure and help apply it to bigger systems and problems. Every time, he came back from the course not only having understood the material, but with ideas about how to apply it in Brompton. And it's plain to see that this is a much better outcome for us than if we had put out advertisements or hired recruiters to get a lean specialist for the pre-assembly department. Craig's learning and development have all been exactly shaped around our needs, rather than those of a different company or a generic example from a textbook. And there's never been any worry about personality or fit; we've known from day one that he's just our kind of person and he's always been committed to Brompton.

Perhaps not everybody has the latent ability to be a lean manufacturing ninja, but everyone can generate ideas. If you have a friend or contact in a different industry from the one you work in, then, in my view, you've got a learning opportunity. Someone in a different industry won't have exactly the same problems as we do, but they may have tackled something quite similar, or something that's not even really similar at all but has some structural analogies that get you thinking. They are more likely to share their ideas with us than our competitors are, and by talking to them we get new ideas which might be part of best practice, rather than benchmarking ourselves against the rest of the industry and then acting surprised when the results are no better than average. I try to arrange regular field trips to other companies in industries not directly related to bicycles: Charles Tyrwhitt shirts, Hayter lawnmowers and things like that. We take a team of four or five Brompton people for a day or two to walk round and ask questions. We can learn what their equivalents are of the socket hinge or the chain pusher plate boss, the little details that tell the big stories.

The principle here is that of making sure you don't accidentally cut yourself off from potential sources of information. The whole problem of management is in some way to match the wild variety of the world with your capability to deal with it. You give yourself a huge advantage in that task by keeping your own mind open, and by encouraging everyone who works for you to use their whole capability and understanding, rather than fitting them into a box. Being constrained to a single optimal path is the right thing for a drill bit in a jig, but people aren't machine tools; you want to multiply their ability to absorb information, not reduce their ability to react to it. If someone brings you an idea and you turn it down, then you can be pretty confident that it will be a while before that person gives you another idea – if they ever do at all. It's personally upsetting to go to the trouble of thinking of something, pluck up your courage to suggest it to the boss, and then have it rejected; only a few very robust people will keep coming back for more of that treatment.

And you can't tell in advance what will work. If my desk had been the kind of place where ideas go to die, I might never have met the woman who invented the nine-day fortnight. I trusted Michelle Mitchell a lot; she had been the office manager in the early days, when Brompton consisted of not much more than a factory floor and a room in which people smoked cigarettes and maintained an Excel spreadsheet to work out the resources. Over the years, and possibly understandably given the environment, she got interested in the business of management and particularly the way in which we tried to encourage good people into supervisory roles and plan careers. After some studying, she became our head of human resources.

One day in 2008, when flexible working was unusual in manufacturing – something one had vaguely heard of software companies doing, perhaps – Michelle pointed out that the official working week at Brompton was forty hours; that is to say, ten eight-hour days every two weeks. This rhythm was set by the shop floor, where everyone had to be in the same place at the same time, but the office staff had to be in on the same schedule (and many of them got in early and stayed too late). If we let the non-production employees have a pattern consisting of nine nine-hour days every two weeks, on the other hand, then every other weekend would be a long weekend for them. For a young workforce based in London, who worked in a bicycle company and loved outdoor activities, this was a substantial advantage. If you wanted to go away at the weekend, you could leave in the lighter Thursday evening traffic rather than the Friday traffic, and get something like a full two days' worth of enjoyment. It turned out to be wildly popular. When the option was offered, around 80 per cent of the office staff took it up.

It's a perfect example of the kind of idea that we try to have at Brompton – it doesn't cost anything, it might make people happier and it's not copying anyone else. More and more companies, particularly since the Covid-19 pandemic, are beginning to see the benefits of not following the crowd, but we were at least ten years ahead of our time. It's also a bit distinctive, which matters because it builds another kind of social system; people identify with the company, they like it and they want it to succeed.

This makes me think that if the culture is right and the company has been set up with systems that generate solutions, attitude matters more than aptitude. Someone who is interested and cares about what we want to achieve is going to do better at Brompton than someone who looks on paper as if they have specific skills. When I look at resumés, it's always the hobbies and interests that my eye is drawn to, because that's where you find the indications that someone thinks for themselves and isn't afraid to be a bit different. If someone's not afraid to be different – in a world where we're all encouraged to be the same – then they're more likely to have a different perspective on what we do, and more likely to have the confidence to suggest doing things in a different way. That is innovation.

Really giving up control to someone you've hired, rather than talking about doing so and immediately starting to second-guess and micromanage, requires a bit of confidence. And self-confidence in management comes from having a consistent understanding of why you're doing what you're doing. I reached mine from an engineering point of view; other people get into business from other backgrounds, but as I see it, developing employees, trust and delegation all come from the same principles as organising a factory. You build systems rather than solving problems – design quality in rather than inspecting defects out – and let people come to you with solutions rather than going to them with instructions. Any time someone asks what they ought to do, you don't answer the question; you try to find out why they had to ask.

GO ANY
“... creates a ne
bicycling becom
FINANCIAL T
...TAKE
ANYWHERE
The Brompton folds smaller than any other bike: yet in use it is a fast, lightweight and versatile machine which offers countless new ways of travelling about.
BR

11 DISTRIBUTION NETWORKS

Ever since Brompton was founded, the company has sold a large proportion of its output overseas. In 2020, 68 per cent was exported, with 30 per cent going to Europe and 38 per cent to the rest of the world.

'Oompala oompala oompala Brompton! Whoooooaaah Brompton! Oh Brompton! Oh Brompton!' Possibly to the tune of an ancient Viking song, this was the anthem of the international Brompton sales gathering, written in about 2008 by Lars-Åke Fredberg of Gamla Stans Cykel in Stockholm. We sang it every year at the distributors' conference, where people came from around the world to be brought up to date with new product developments, share views and ideas on marketing and talk about what was working and what wasn't. (And also occasionally to give me a hard time for not being able to send them as many bikes as they wanted.) I will always associate that remarkable song with a particular period in the company's development and my life. Things aren't quite the same now; the distribution model has changed and the Oompala song isn't heard as often as it used to be. The reasons why things had to change all related to money and control, and the relationship between the two.

The changes were also driven by the way in which that relationship changes at different scales of operation. So perhaps the best way to think about them is to start at the beginning and move forward through time, looking at how Brompton has had to keep adapting the way it deals with getting the bike from the factory floor to someone who wants to buy it. We'll touch on marketing and branding, but let's keep the abstract and philosophical questions of 'what makes the brand' for later. Right now, just understanding the physical and logistical operations of maintaining a dealer network is enough to think about.

We'll start with some definitions. The supply chain for bicycles is made up of manufacturers, retailers and distributors. Manufacturers are simple enough to understand; that's us. (It's a bit more complicated when spare parts aren't made by the same company as the bikes they fit onto, but essentially it's simple enough.) A retailer, or dealer, is the shop that people walk into; also straightforward. And a distributor is the company in between the two. Distributors hold stocks of bicycles and supply the retailers, but also have a few more responsibilities. They produce catalogues, co-ordinate retailer training and check that standards are maintained for displays, spare parts availability and so on. Things are perhaps not quite so clear cut in real life (some distributors are also retailers), but in theory at least, these are the three roles.

In the early 2000s, when I had just joined the company, there was not much thought given to the structure of how distribution and sales were handled. Hardly any was needed, given the scale that we were working at. Our marketing effort largely consisted of putting together a brochure and making sure that somebody answered the phone when a bike dealer called up to place an order. For the domestic market in the UK, we were our own distributor. People used to see the bikes and like them, and then ask their local bike shop how they could get one, and over time these enquiries led to us building up a network.

It was particularly important to us to get these relationships right. The Brompton can be used for fun, but it was designed as a tool for urban living. If someone is using it for their daily mobility, according to Andrew's original vision, it really matters to them to be able to get it serviced or repaired quickly. Although the overall life of the

One of our first Brompton dealer days: Brentford, 2004.

bike is determined by the fatigue life of the big metal parts, smaller parts are subject to wear and can give up at unpredictable intervals.

If you can drop your bike off in the morning at the place where you bought it and pick it up again in the evening, your life isn't that disrupted – if you can get it serviced over a weekend and have the parts replaced before they fail, even better. On the other hand, if you're left for a week or so without your main means of getting to work, you're going to feel that the original promise made to you has been broken. And it's only human nature that any bad feeling will stick around.

This means that from the company's very earliest days we've regarded the sale of a bike as only the start of our relationship with our customer, and we've emphasised that every shop selling Bromptons needs to be able to follow through. They're not fundamentally toys, and they're not designed to be grown out of; it's important to the brand that we design obsolescence out, not in. The Brompton is an unusual product, with a folding mechanism that has its own stresses and wear and a lot of unique components. So people need training in order to be able to maintain it. And in order to provide the service that we expect, the retailer needs to keep a good stock of spare parts, and to have a mechanic on hand who has been trained specifically on the Brompton.

Organising this was challenging enough in our domestic market. It would have been enormously more difficult to establish and look after an overseas dealer network, working in foreign languages, different timezones and countries where we didn't know the market. Even things like finding a list of bicycle shops and knowing which ones were located near to each other, which the internet now makes simple enough, used to be more of a challenge. In order to drive export sales, Brompton had therefore needed to find overseas distributors. In any given market, we would agree with one company that they would be the only importer of Brompton bikes, and in return they would take on a lot of the support roles and handle marketing and training.

The usual Catch-22 applies here; if you're the kind of international brand that could well handle the whole thing yourself, distributors will be lining up to take you on, while if you're a smaller operation it's much more difficult to attract interest. Andrew was not a great one for hustling, so in the company's early days Brompton tended to open up new overseas markets in the same way that we added domestic dealers – people heard about the bike and liked it, then approached him at a trade show or called him up.

Lots of cycle manufacturers try to generate a party atmosphere at trade shows, with free booze and attractive sales staff, almost as if the aim was to divert attention away from the engineering. This has never been our style. To this day, the Brompton stand at a trade show is not flashy or particularly impressive; you'll find one or two employees (who have usually assembled the stand themselves the night before with a spanner and some Allen keys), no free food and drinks, and a few demonstration bikes. We want the bike to be what draws the crowds, not the stand, so we can talk about the product and what it can do for your life.

This may seem like an old-fashioned and small-time way to operate. A modern startup would probably not bother with trade shows at all: they could just sell a stake in the company to a venture capitalist at a startling valuation, then use half the money for a massive global ad campaign and the other half to buy fulfilment services from an outsourced provider. But the Brompton approach was completely appropriate for us. Our international distributor network grew slowly, but every single distributor we signed up in the early 2000s was led by a wonderful person who believed in the product. We had people like Hans Voss, a German plumbing supplies distributor who once visited a bike show practically on a whim, bumped into Andrew and developed a sideline in distributing folding bikes that outgrew the rest of his business. There was Channell Wasson, a war veteran with a handlebar moustache who could have been the prototype for every bike hipster in the USA. He

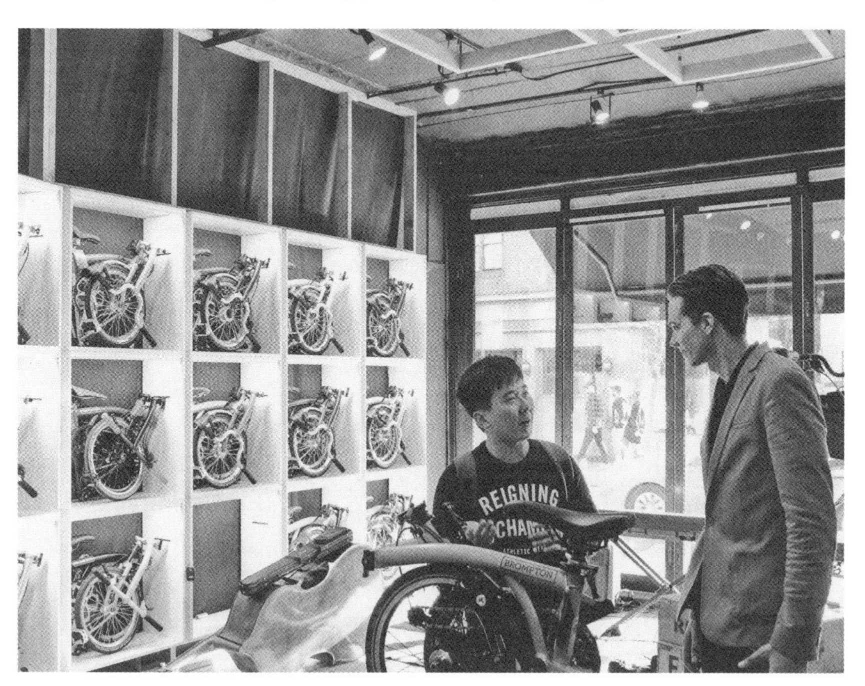

Getting ready for the opening of our NY Junction in Greenwich Village, Manhattan, 2018.

lived in Palo Alto, California, and he used to get us to ship over huge solid rubber tyres for his penny-farthing from the last company in the world that made them. It always amused me that Channell was involved with both the small-wheeled Brompton and some of the largest bicycle wheels ever made, but that was the kind of person who used to be drawn to us. And then there was Simon Koorn.

Simon is a larger-than-life character in every way – he's a huge man, even by the standards of the Dutch. He also had a huge heart and a huge love for the Brompton bike. And he was a key distributor, particularly early on. He had been importing Bromptons into the Netherlands since 1988 – basically, from the very start of serious and sustainable manufacturing. He had been making suggestions about the design of the bike from the beginning, and he regarded himself, with some justification, as having had an influence on its development. Everywhere else in the world, we had standard model codes – the letter S, M or P for the handlebar style, a number describing the gearing, and another letter saying whether the back of the bike was empty, fitted with mudguards or had a rear rack. In the Benelux region, on the other hand, the models had names. At some point, Simon had become fascinated with the fact that Brompton was a place name, and had apparently looked through a road atlas and a dictionary of names to find other words which had an appealing ring to Dutch and Flemish ears. So his 2005 catalogue had a 'Brompton-Regis', a 'Brompton-Ralph' and the 'Patrick-Brompton' and 'Potter-Brompton'.

This sort of thing was bound to cause problems in the long term, and it did, but to start with it was a real source of strength to have so many wonderful characters involved. The Brompton had been designed by Andrew for his own particular requirements, and although we all believed that lots of other people would get value from it too, that didn't mean we had a coherent vision of how to market it. We learned a lot from the distributors; every one of them had ideas. It was like having a laboratory in which we were always carrying out experiments and finding out what worked best.

As we added distributors during my first few years at Brompton, it was a case of 'the more the merrier'. I had noticed early on that Andrew didn't really like to travel, so I thought I might make myself useful by going out and meeting people, trying to make more contacts and signing up more distributors. That was how I ended up arguing at the Taipei International Cycle Show with a man who collected samurai swords.

'If we don't get these problems solved, we're going to find another distributor.' I knew I wasn't supposed to talk to Toshi Mizutani like that. He was a Japanese businessman of the old school, thirty years older than me, with stiffly upright bearing, neat grey moustache and wire-rimmed glasses: the complete package. Plus he was president of the Japan Bicycle Association and the proud representative of his family's company. For someone like me at the age of twenty-nine to walk up to an executive of his age and status in public at a trade show and start berating him about business problems was more like sacrilege than mere bad manners. And initially, he seemed as much incredulous as angry.

But good people don't stay angry about being told the truth. Mizutani Bicycle were (and are) a fine company. They started as a manufacturer in 1924, but by the 1990s they had given up making bikes in order to concentrate on distribution. As you'd expect from a company run by a man like Toshi Mizutani, they were great at paying on time, great at keeping their promises and impeccably reliable for logistics and dealer training. From Brompton's point of view, though, the relationship was a disaster.

The problem was that at the time they sold a few real Bromptons, but they also sold Neobikes, the disastrous Taiwanese 'not-Bromptons' produced under the ten-year joint-venture licensing agreement with Eurotai. This was early in my career, when the agreement was coming to an end, and they were a sore point; the bikes were really bad, and they were being sold next to the real thing. But if we were going to replace Neobike sales with the real thing, the distributor economics would be badly screwed up. That's because Mizutani imported Brompton bikes and spares through a trading house and sold them on to Japanese retail. Everyone up and down the supply chain was taking a profit margin; it was never clear quite who was responsible for the end price, but in the very early days of our presence in the market, Brompton bikes were crazily more expensive in the shops in Japan than anywhere else in the world.

This didn't just mean sales being lost and a niche being created for clones and copies to creep into – it was beginning to generate bad feeling of the sort that lingers. Japanese Brompton owners were increasingly getting access to international price lists from the internet and they could see that the bike was selling for a lot less in the West. Not unreasonably, they resented this. Understandably their resentment was directed at Brompton rather than any of the other links in the supply chain, and it was pretty clear that unless

At Bangkok central rail station in 2018, after cycling across the city with my daughters and our fantastically enthusiastic Brompton dealers. It was very warm and very wet.

the problem was addressed it would only get worse. We would never have been able to build up a relationship with the Japanese consumer which has served us so well if we had begun by making them feel they were being taken for a ride.

These were quite sharp points to be making in public to Mr Mizutani. Word got back to London pretty quickly and words like 'lunatic' and 'loose cannon' started to be thrown around. Tim Guinness had to add another stop on his next business trip to restore the peace. Solving the problem, though, once everyone had been forced to admit that it existed, ended up being surprisingly easy; just a bit of time on logistics and paperwork made it possible to remove links from the chain, and cutting out the profit margin on those interim stages made it possible for Brompton, Mizutani and the retailers to earn a living while selling the bike at a more reasonable price.

The 'Battle of Taipei International Cycle Show' might not feature prominently in promotional videos or cultural awareness courses. But far too often, trying to handle things sensitively means not handling them at all. The relationship was a lot better served by telling Mizutani the facts, to his face and in short, sharp sentences, than by pretending or hoping that all the problems would go away. A few years later, Toshi Mizutani invited me and Quinton Pullinger, our head of Asia, to join him on a visit to the antiquarian armoury where he bought his swords. It's the kind of shop where dilettantes and outsiders are not normally welcome. Slightly hungover, I started reaching out to check the edge on one of the blades. Mr Mizutani swiftly held me back, with a warning that this was a quick way to lose a fingertip. If he had still been sore about the argument, that would have been an easy way to get revenge.

Sorting out the relationship in Japan was a huge deal for Brompton. Japan is a logical first point of entry into the whole of Asia for a Western manufacturer – it's the biggest market in Asia, it's got a solid legal system and commercial culture, and, although they might not like to admit it, consumers in many other Asian markets tend to follow Japanese fashions. By 2008, we had distribution in Taiwan and Singapore and were sending more and more of our production to Asia.

This wasn't actually very good for profitability. Every bike we sold outside the UK reduced our margins at the time, because only in the UK were we taking the distributor's profit as well as the manufacturer's. Also, because sales were limited by our output, every bike shipped to an overseas distributor was one less for British dealers. But we needed to do it, for the simple reason that – again – you can't plant seeds at harvest time. We were investing in capacity to increase our output, and we were concerned (at the time) that the UK market was going to become saturated. From today's perspective it seems strange to have worried, but we really didn't know how many people there were with the right kind of transport needs who were willing to consider a Brompton as a solution. So there was a danger, when looking for new distributors between 2002 and 2005, that we wouldn't have enough orders for the bikes we were planning to make between 2006 and 2010. It takes at least two years and often as many as four from the start of a new distributorship before the market is really ready to take off. A large part of managing the growth of a business is understanding the lag between when you act and when you start to see the effect, and taking this lag into account in planning ahead.

By 2014 I realised that, having invested all this effort in building a network of distributors, we were going to need to change the model; if anything, we might have left this decision a bit late. We had some amazing distributors – Lars-Åke and Hans (both of whom had sons who had followed them into the business), Koos Kroon, who effectively introduced urban cycling to Barcelona, and many others. But they were enthusiasts and in many cases small businesspeople who relied on their distributorship profits for an income. These owner-managed firms would have found it difficult to make a big investment in growth; even a relatively quick payback period would have left them with nothing to live on for a year or two.

And because of the internet the world was becoming a smaller place. People were comparing prices and products across different territories, and we needed consistency. Nor was it still viable for every bike configuration to have a different name in Benelux. Gradually we started to take more control of the distributor network, initially by persuading everyone to sign up to formal contractual agreements, but eventually by buying them out and taking full ownership of our international sales.

This was neither easy nor pleasant. The law relating to distributors is quite complicated, and it also varies considerably from country to country. The basic principle is that the relationship between a supplier and a distributor is different from someone buying steel tubes or grain on the wholesale market. Distributors are expected to make their own investment in the product; they build their business around it. It's not fair if someone spends years building a brand, getting a product into retailers and supporting it, and then has it all taken away just like that. The law on this is not like employment law, but nor is it completely unlike employment law. One feature it has in common with much employment law is that the less that's written down (and this is particularly true in southern Europe), the more rights the distributor has.

We started the process by making an offer for the Brompton distributorship of Fiets a Parts, Simon Koorn's business in the Netherlands. Dutch law isn't particularly favourable to distributors, and the written distributorship agreement was very detailed. We also felt that with this distributorship in particular, however much we loved Simon personally, we really had to do something. We met in Brussels for one of the most agonising and emotional meetings I've ever held.

Our offer was significantly more than the legal minimum; it wasn't as much as Simon's estimate of his contribution to the Brompton story, but it couldn't have been; Brompton had been his life for years, just as much as it was mine. And it was a lot of money, enough for us to need two years to pay it. We wanted to treat him fairly; as well as the personal relationship, we had to remember that we were still a distributor-driven business at the time. We had to think about how the way we dealt with Simon was going to be seen by our other distributors. Although most of them agreed that things couldn't go on as they were with Simon, and that a change was needed, people liked him, and they would see this deal as indicative of how we might treat them in future. We couldn't send a message to all our other partners that if you poured effort and investment into Brompton over several decades, you could end up having it taken away with the bare legal minimum payment. I hope people understood this, but as far as my relationship with Simon was concerned, the damage was done. Two weeks later I walked into his office in Groningen. The documents were ready. Reading and signing them took about thirty minutes. And that was that – there wasn't anything left to say.

Negotiating these deals took several years and a lot of effort; for better or worse, it may have distracted our attention from improving the bike during that time. None of the deals was quite as harrowing as the experience with Fiets a Parts, but none was particularly easy either. The nature of the transaction is that you're taking either all or part of someone's business, and putting a single number down on the table as the cash valuation. That's the commercial reality – money and control – but it is quite difficult to face up to it sometimes. It would have been so much easier to stay in a comfort zone, singing the Oompala song together and acting like we were just a group of international friends who happened to sell the same make of bike, but that wasn't what the good of the company required.

At the end of it, though, the company had benefited substantially. We had higher margins and more control. We were also able to sell through a wider variety of channels. Our original distributor agreements had usually been exclusive, so we couldn't sell to larger chains, or directly to consumers, but now we could. This became more and more important as we started to launch our own network of retail bike dealers.

We had begun doing this even before buying out the European distributors, in 2011 and in a territory where we still continue to sell through a distributor – in Toshi Mizutani's homeland, Japan. Hiroshi Oeshi of Loro Cycles approached us with an idea for a 'concept shop'. He is a more modern style of businessman, and a real enthusiast for slightly offbeat bicycles. His stores specialise in not only folding bikes but also recumbent ones. These have a dedicated community of aficionados who mostly recognise that the quirkiness is part of the appeal as much as the allegedly superior ergonomics. Oeshi has grey hair too, but the physique of a keen recreational cyclist. He loves bicycles for the life that they can provide.

His idea was that Loro Cycles would open a dedicated shop called 'Brompton Concept Store Kobe Loro' in Kobe, Japan's seventh-largest city. This apparently rolled off the tongue a bit better in Japanese than in English. We didn't want to call the stores just 'Brompton', so as to keep that name special for the bike, but we had only two weeks to come up with a better name, and everyone in the company started furiously brainstorming. Shortly before the deadline, our Asia manager, Quinton Pullinger, came up with the winning idea. If the aim was to bring Brompton enthusiasts together, then we should name the stores after a railway interchange where people come together. The largest such interchange in London is Clapham Junction, so the stores would be called Brompton Junction. (We even tried, briefly, to extend this concept to a logo based on the London Underground map.)

Three weeks before the scheduled opening of Brompton Junction in Kobe, the Fukushima nuclear power station was struck by the 2011 earthquake and tsunami. Everyone presumed that the launch would be called off – expats were evacuating from as far away as Hong Kong and telling me that it was crazy even to consider going to Japan. I took a deep breath and looked at a map. Japan is a big, narrow island and Fukushima is about 500 kilometres north of Kobe. The winds were blowing east–west; it seemed to me that there wasn't actually any danger of radiation. So I called Oeshi and

Having control of our own retail space allows Brompton to show off new models to their best effect, as seen here at our store in Covent Garden, London.

confirmed that I would be there. Apparently this significantly raised my status in the eyes of all our Japanese customers, for standing by them at a time when the whole country was understandably feeling traumatised.

Brompton Junction in Kobe was a success from the start, and we've continued to open up flagship shops ever since. This is a skill set that's way outside my comfort zone, but we were fortunate in having hired Lorne Vary as CFO. He came to us in 2009 from the food company Planet Organic, and he thoroughly understands retail. It's fairly unusual for the finance director to be in charge of a sales channel, but as we saw earlier, it ought to be much more natural for people to step into roles where they have ideas and passion, rather than being recruited into a particular spot and never moving outside it.

The Junction stores are flagships as much as retailers. We don't have any ambition to put one in every town; instead, the idea is to have a big storefront in an aspirational location that's dedicated to presenting Bromptons to their utmost advantage. In the early days of Brompton, we tended to get pushed into the back of the store. In our own shops, people can see more of the range of available colours and begin to visualise themselves on the bike. They might not always walk out with a Brommie, but the idea gets planted; maybe they buy one six months later from their local shop. So we're creating a system which uses a sales channel to drive sales. Rather than have bikes hanging around in our warehouse, or the warehouse of a distributor, we put them in a place where people can see them and buy them.

Furthermore, the stores help show all the other bike retailers that we deal with how to do a great job of selling our product. We can use the Junctions to innovate and experiment, and we then share what works with the whole dealer network. This might mean coming up with new promotional ideas, or it might just mean trying to make the whole experience more like a normal retail shop, rather than the stereotype of a shed packed full of bikes, with an all-male staff who look down on you if you don't go racing at the weekend.

HRH Prince Charles watches a demonstration of how to fold a Brompton at Glasgow Central Station.

In 2013, we wanted to open a store in London, but nobody would let us. The prestige locations where we wanted to put ourselves actually refused our money – they wanted to maintain their image with premium fashion brands, not bike shops. Finally a friendly estate agent gave us a tip about a last-minute opportunity, and Lorne and I rushed into central London to buy a bankrupt Belgian chocolatier from its liquidators, just because it was still the tenant of no. 69 Long Acre in Covent Garden and that was the only way of obtaining the lease. It was an amazing space, but the rent was far too high; about £20,000 more than the budget which the board had approved. Lorne and I walked across the road to think about what we were going to do, and as we turned to look back at the building we saw a green plaque on it, featuring a man in a tailcoat riding a bicycle. It read 'DENIS JOHNSON (1760–1833) from his workshop on this site in 1819 made and sold Britain's first bicycle in its Hobby-Horse form.' That made our minds up. It had to be; we found a way to make it work, and took the lease. Five years later, when we opened in Greenwich Village in New York, everyone was happy to be associated with us.

I've brought this story a long way now from fillet brazed joints, malleable iron castings and chain pusher plates. But there's a sort of intellectual consistency to the way I've tried to run Brompton as CEO. When you move from production into management the problems get more varied, but the logic is surprisingly similar. Everything stems from tolerance and variation: how far the real world is capable of diverging from what you think it ought to do, how much it's possible to stop it from doing so, and the allowances you have to make for the possibility of something unexpected happening. Complexity, control, organisation, influence and relationships; building a company is about creating an environment in which people generate solutions and ideas.

But the purpose of life isn't to make blood circulate around your body; that's back to front. Building a company is only worthwhile if the company is going to be viable and able to develop – but it's got to be *for* something. After talking a lot about solving problems or trying to make them solve themselves, it's time to look at what we actually want to do in the world.

PART 3

CHANGING THE WORLD

OMPTON Order form

OMPTON BICYCLE (SALES) LTD.,
e Old Powerhouse,
w Gardens Station,
hmond, Surrey.

ase supply:

n	Quantity	Price exc.VAT
OMPTON BICYCLE	_______	£170
namo lighting	_______	£11
ont luggage carrier	_______	£10
ver	_______	£20

nderstand the terms relating to this order, and I lose a remittance of £25 as deposit in respect of h bicycle ordered.

ne:

dress:

tel:

ned:

e:

Complete and detach the form on the left and send it to us together with your deposit.
(If you are ordering after the date given at the foot of this page, please contact us to check prices.)

BROMPTON SPECIFICATION

The BROMPTON is a full size, 16" wheel bicycle which folds easily into a package 20" x 22" x 10". It includes the following features as standard:-

* *3-speed gears*
* *rear luggage carrier & shock cord*
* *swinging arm rear suspension*
* *folding crank for left hand pedal*
* *full size pump*
* *castors on the rear carrier*
* *fully adjustable saddle height*

EXTRAS

Note:- these do not affect the ease of folding or the overall size of the folded package.

DYNAMO LIGHTING: *the dynamo is attached to the rear frame and is easily engaged or disengaged; the wiring is enclosed for protection, and the system provides first class lighting.*

FRONT LUGGAGE CARRIER: *this comes complete with shock cord, and provides a useful platform 10" x 5" for extra luggage*

COVER: *smart but tough canvas cover which slips easily onto the folded package, with a flap which may be tied across the base; this fully encloses the bicycle*

TERMS

1. A deposit of £25 per bicycle should accompany each order - except as stated below, this is non-returnable

2. Bicycles will normally be ready within 4 to 8 weeks from the date of ordering. Where delivery is likely to exceed 8 weeks, we will advise you of the expected date of completion. If you notify Brompton Bicycle (Sales) Ltd. that this is not acceptable, then, providing we hear from you more than 6 weeks before the advised delivery date, your order will be cancelled and we will refund your deposit

3. Payment of the outstanding amount on your order (total order value + VAT, less deposit) becomes due as soon as you are informed that your order is ready. If payment is not made within two weeks of your being so notified, delivery of your order may be delayed

4. Your order, when ready, may be collected from Kew, from an authorised agent or as outlined below (London Delivery). If a special arrangement has to be made, then a charge may be made.

5. Brompton Bicycle (Sales) may alter the detail specification of the BROMPTON.

LONDON DELIVERY: If you live or work within 5 miles of Kew or Hyde Park Corner, we will deliver to you by appointment, and at the same time demonstrate and give you full instructions on using your BROMPTON. If you would like to use this very convenient service, please indicate on your order form - the charge for 'London Delivery' is £8.50.

PRICES:

THE BROMPTON

BROMPTON BICYCLE (SALES) LTD.,
THE OLD POWERHOUSE, KEW GARDENS STATION, RICHMOND, SURREY
Telephone: (01) 940 2879

We ended the last section by talking about sales and distribution networks, getting close to the subject of marketing but never quite touching on the subject of the Brompton brand. That might have seemed like the logical next step. But it isn't. A brand isn't a thing you do or make; it's not something separate from the company. A cool person doesn't spend a couple of hours every day working at being cool, it's something that arises naturally from the other things they do. They might have a publicist who tells people about what they do, and makes sure that they're communicating with the world, but that's not the same thing.

We are not in the business of consumerism. Our products are not gadgets or fashion accessories; they're tools which people can use to change the way they live in cities. We don't try to build in obsolescence to force customers to buy more; we take pride in designing things which will last as long as the physical integrity of the metal they're made from. None of the weight of a Brompton is ever going to come from material that has been added for styling purposes. As the 'Brompton Philosophy' document that we published in 2016 says in large type: 'Brompton is Honest'.

So when we talk about 'branding', we really mean the relationship that we have with the consumer. The Brompton isn't an obvious solution to anyone's problems; people have to be brought to it. That sometimes means partnering with other people and organisations who share our values, and it sometimes means trying to tell stories that people understand and which lead them to the bike. A lot of the time, the brand is built on word of mouth, and that comes about by supporting and valuing our customers. If we keep doing this, we should continue to grow.

HRH Prince William giving the Brompton a whirl at the Great British exposition in Shanghai, China, in 2015.

I left ICI because I wanted to have an effect on the world, and I didn't think I could achieve that within an organisation that was so much bigger than any one person. When I joined Brompton, I thought I would learn and develop for just a short time, and then move on to do something else that really made a difference. That's not what happened, of course – I fell in love with the bike and the company and stayed – but as time has passed and I have grown up, it's become more and more obvious that human-powered transport could change the twenty-first-century world as much as any other industry, and more than most.

We should be judged by the difference between the effect that Brompton has on the world, and the effect that it could have. The company has always been ambitious, but for a long time we didn't have a clearer idea of our potential than that it had to be bigger than ten or twenty thousand bikes a year. For years, a magnum of champagne sat under first one then another of the many different desks I had; it had been presented by Professor Tim Baines from Cranfield University, after he brought his students into Brompton for a consulting project that changed the way we saw the company. The gift was meant to motivate me to get to 50,000 bikes a year. That was a huge challenge which had taken fifteen years to achieve by the time we finally opened the bottle at a fantastic Brompton celebration in 2019 (we also bought a Nebuchadnezzar of Bollinger with fifteen litres in it, as we had a lot more glasses to fill by then). But now our ambitions are much greater, and that's slightly scary. Going from 5,000 bikes a year to 50,000 required massive changes to the factory, the company and the mindset. Further order-of-magnitude growth will hardly be any less of a challenge. But it's what we need to do.

The world needs bicycles, and light electric vehicles. Most of the world's population live in cities, and the proportion doing so is only going to grow. But cities can't go on the way they are. As currently designed, the late twentieth-century developed-world city is a machine for destroying the mental and physical health of its inhabitants, while also contributing to the world's environmental crisis by pumping out pollution and greenhouse gases. The best and most serious thinkers about cities all believe that cycling – or perhaps more accurately, a return to cycling – as the predominant mode of urban transport is the only viable option for the future. The Brompton has the added advantage that it can fold, is compact and light, can be stored anywhere and combines with other modes of transport. That makes it perfect for city life.

But you can't change the world even by making 50,000 bikes a year. We need to set our sights a lot higher. I've mentioned that part of the job of building a company is to make sure people can create viable, ordered systems. Increasingly, though, I think another part of my job is to create disorder in the right places and quantities. Because if the company just keeps doing the same thing, it will keep getting the same results, in the same way that if you benchmark yourself to the rest of the industry, you will never rise above the average. I need to create a vision and ambition, and then push people to do things differently in order to get there. So the next question is, how are we going to push out of the comfort zone, creating trouble for my long-suffering employees so that they find new ways to excel?

18V

12 THE PROBLEMS WE NEED TO SOLVE

In 1800, only 2 per cent of the world's population lived in cities. By 2016, that proportion had risen to 54 per cent. According to the World Bank, cities account for more than 80 per cent of the world's economic output and 60 per cent of greenhouse gas emissions.

I didn't grasp it straight away, but travelling around the world and learning how the Brompton had changed customers' lives gradually helped me to understand the purpose of the company. The secret to the Brompton is that everything about it is different. Which means that you stand out in public when you're riding one. Not everyone likes that, which means that Brompton owners are always slightly unusual. But they tend to be unusual in one of two distinct ways.

The first group of owners are probably the people that Andrew had in mind when designing a folding bike that he could use himself. Whether literally or not, they're engineering types. People who see things in functional terms and don't care too much about aesthetics. They tend to get the concept very quickly: they understand the benefit of the folding mechanism and the light weight. They are OK with riding a Brompton around in public because they generally aren't thinking about how they look to other people; they're thinking about what they want to do.

The second kind of Brompton rider, though, is different. These are people who do care about aesthetics, and that is why they like the Brompton. They aren't indifferent to how they look when they're out and about, but they are confident about their own sense of style and design. These people understand that looking different can be good. So they aren't worried about looking different because they're riding a folding bike – they know that they are, in their own way, cool, and therefore the bike must also be cool because they're riding it. A lot of the reason why the bike does so well from word-of-mouth recommendation is that people who own Bromptons like to talk about them, and the kind of people who might want one are more likely than most to ask the owner questions when they see one being ridden.

Over the last twenty years, hardly any Brompton owner I've met has said anything other than that the bike has changed their lives. And I could see that the owners have something in common. They have a particular way of seeing the world; they don't mind being different. And they're a bit less risk-averse than the median person; unfortunately, you still have to be in order to cycle in a city. But they're happier because they own one, and healthier. Based on everything I've seen, there's no way we could be content in leaving Brompton as a niche, cult product with a small fan base of enthusiasts.

One other thing about that fan base, though. People sometimes say the Brompton is an 'architect's bike'. That's something of a stereotype, and it's no more true than saying some two-seater sports cars are for hairdressers. But the thing that makes any stereotype wounding is always the grain of truth within it, and of course architects are exactly in the intersection of the two kinds of Brompton enthusiasts; they are engineers who understand the relation between form

Another happy customer.

and function, but they're also supremely confident about their eye for design. It turns out that a lot of them are also very open to new experience and have more risk tolerance than the average. And they understand how urban spaces work. That's why so many of them really do buy Bromptons.

Going around the world and talking to local architects on bike rides is enjoyable in itself, but it's also a great way to learn about cities and the way they are put together. The Danish architect Jan Gehl talks about 'the human scale', which is also the title of an excellent documentary film about him, his architectural practice and his ideas on urban planning. Those ideas are based in part on lessons learned from the modernisation of the city of Copenhagen, where conscious decisions were taken to build the city around the people who live in it and to make the city centre into a liveable place with public space. In Britain, similar ideas have been championed by Peter Murray, the architect and writer who chaired a report on cycling in London for the then mayor Boris Johnson. Both Murray and Gehl's advocacy of cycling comes as the conclusion of a set of arguments, not the starting point. The starting point is the simple but revolutionary idea that people need to take their public spaces back.

And specifically, they need to take those spaces back from cars. Transport infrastructure is one of the central problems of urban design. But even as far back as the 1970s, people who were thinking seriously about cities were able to see something that almost everybody in the bicycle industry had forgotten and only Andrew Ritchie was beginning to remember – that cycling is a mode of transport, not just a sport.

In 1898, the first-ever international urban planning conference was held in New York. One key problem that the planners debated was the 'horse manure crisis'. At that time, there were about 75,000 horses on the streets of London, pulling carts, drays and hansom cabs; other world cities had similar equine populations. A working horse eats 10–15 kilograms of food per day, and produces roughly the same amount of excrement, plus a couple of litres of urine. So the streets of London at the turn of the century needed to be cleared of roughly a thousand tonnes of horse pollution every day. It attracted flies and spread disease. This was a problem that was bringing major cities to the brink of crisis; in London, *The Times* predicted that 'in fifty years, every street will be buried under nine feet of manure'.

As we know, the problem was resolved by the invention of the internal combustion engine. But the trade-off was that the city had to be given over to cars and in many cases extensively redesigned for them. That's not an exaggeration. Setting aside all the incentives and hidden subsidies which are given to motorists, it's possible simply to see the effects on a satellite photo. Whether the measurements are done by advocacy groups or property consultancies, the results are pretty much the same. Even in a city like London, which for the most part still maintains its historic street plan, between 15 and 20 per cent of the above-ground surface area of the city is taken up by parking spaces. Most of these spaces are on the streets, where residents are allowed to take up a bit less than twenty square metres, for a relatively low annual charge. That's in a city where the average value of industrial land is £5 million per hectare; residential land ranges from £7 million to £90 million. The city effectively provides every car owner with an asset worth at least £10,000, and possibly several times that, for a charge of barely 1 or 2 per cent a year.

That's just the parking. When you add in the actual roads, then even a city like Paris has 30 per cent of its area dedicated to spaces that are designed around cars and barely accessible to anyone else. In American cities, the proportion can easily reach 40 or even 50 per cent. Even if we ignore questions of pollution and global warming – which we absolutely cannot afford to do – cars in cities have reached the same point which horse-drawn transport had reached at the end of the last century. If we carry on as we are, we are going to end up – figuratively speaking – in at least nine feet of shit.

This is the sort of issue that Peter Murray and Jan Gehl have been warning urban planners about for years. The greatest demographic trend of the post-war era is that more and more of the human race lives in cities. This has happened under all kinds of political and economic systems; it's one of the most easily verifiable and generally observable facts in geography. A major problem of the twenty-first century is how to manage urban living as a result of this growth.

This applies all over the world; cities are more similar than they are different. European cities have problems from being constrained to a medieval layout, American cities have problems from having been designed on an incredibly car-centric grid system, and Asian cities often have problems from twenty years of rapid and largely unplanned growth. But the local and particular difficulties are about 10 per cent of the issues of urbanisation – the other 90 per cent are simply the consequences of the same problem, of getting so many people to live in proximity to each other, bringing in enough food for them to eat every day and moving them from where they sleep to where they work without bumping into one another too much. A city is a huge factory; like the factories that I'm more familiar with, it needs to be organised as a series of systems which generate the solutions to their own problems rather than needing constant intervention from above. And, again like a factory, its problems change qualitatively with scale; sometimes as you grow or respond to change, a complete reorganisation of the space and flow of activity is needed.

An intrigued young cyclist in India admiring the peculiar folding bike.

This does actually happen from time to time. The car solved the problem of the horse manure crisis by reorganising the city, trading off the use of space against one kind of pollution. But cities were less densely populated in 1898. If they could have afforded one, and ignoring for a moment the manure problem, there was no particular reason why every household in 1898 shouldn't have had their own carriage parked outside on the street. There would still have been room for children to play and for pavements to be usable, and road traffic would be an occasional inconvenience. That was even how things were in the early days of the automobile.

But as things grow, they interact and get in the way of each other. At peak times, in congested traffic, a shocking 30 per cent of cars on the road are cruising around looking for a parking space. This is close to a general law, which has been found in studies dating back to the 1930s. What has changed is that 'peak times and congested traffic' account for a larger proportion of the day and a larger area of the city with every passing year. And over the last decade, the arrival of Uber and the like means that it's not just licensed taxis which are driving around empty looking for a fare. On a bad day, that can account for a further 5 or even 10 per cent of the traffic on the road.

So we arrive at the point where, for a lot of the day in city centres, as much as 40 per cent of car traffic is engaged in activity which has no value, and is being driven by people who don't want to be doing what they are doing. At the same time, 90 per cent of residents don't use a car on any given day; nearly half of households in London don't even own one. But nonetheless, they have to give up space which would be worth billions of pounds at anything close to market prices. The car industry was a boon to mankind maybe fifty or sixty years ago. But more recently, at least in the cities, it's been a politically astute lobby group, extracting a huge subsidy and making things worse. The people need their space back.

So far, very little that I've said has related to the inherent problems of the internal combustion engine. But electric cars are not really part of the solution. If we're going to treat batteries and motor magnets with the respect that they deserve, it's crazy to continue the main delusion of the petroleum age – that a 100 kg human who can easily be carried on a 15 kg bicycle (an electric Brompton, say) needs

You can fit forty-two Bromptons into a single car-parking space.

to have a two-tonne car to sit in, taking up city road space which could accommodate another ten humans. A car, even a Tesla, is engineered so that people inside it can survive a crash on a motorway at 80 mph; there's no need to have such a big metal roll cage to protect you when you're travelling at reasonable speeds in the city. The mathematics is wrong, by orders of magnitude.

However green the generation of electricity may be, for personal transport it needs to be stored in batteries. Modern batteries are made using lithium and cobalt. Lithium mining is bad enough. Either you pump huge amounts of water (around 500,000 gallons per tonne) into salt flats in Chile and Argentina, or you dig ore out of the ground in Australia and crush it with sulphuric acid. In the first case, you're transporting huge amounts of water into one of the driest places on earth and then polluting it; in the second case, you're creating huge amounts of toxic run-off. And even adding up all the ore known to be in the ground and all the stocks of metal wherever they are held, it's far from clear whether the world's supply of lithium is enough to make all the batteries that we want. In principle, it's possible to extract lithium from sea water, but the process is hugely energy-intensive.

Cobalt mining is, if anything, worse. It takes place for the most part in the Democratic Republic of Congo. With most mining operations, the environmental problems arise as a result of the ore being dug out of the ground and then treated with horribly toxic chemicals to remove the metal from the rock and other elements that it's found in compound with. Cobalt is different: it can be dug out of the ground in a reasonably pure form, but the metal itself is toxic. Unfortunately, the fact that it doesn't need large-scale industrial treatment means that people can set up 'artisanal' cobalt mines and dig for it at small scale, without any functional worker protection or environmental regulation.

The problem with electric cars isn't a question of 'charging infrastructure' – it's a simple matter of physics that there just isn't an efficient way to lug a two-tonne steel box around with you everywhere you want to go. It's an equally simple point of geography and urbanism that twenty bikes can fit in a single parking space and that six cyclists take up the same road space as a single car. The battery in an Audi Q4 could be divided between 150 e-bikes, for which the charging infrastructure already exists. Putting in car charging points is a sign of a city government that's not got the nerve to address its real problems but wants to feel like it's doing something vaguely green. The bicycle is the most energy-efficient form of transport, so the electric bicycle is the most efficient mode of electric transport.

That isn't to minimise the problems caused by the internal combustion engine; if the underlying problem is the car itself, the engine is what makes it urgent. In London – which is by no means the worst city in Europe – it's been estimated that over 4,000 people die young every year because of the air quality. The city isn't compliant with the EU minimum standards for fine particulate matter in the air, which are themselves twice as lenient as the WHO standards. This has lifelong consequences for children. Ironically, and perhaps surprisingly, cyclists and pedestrians don't suffer as much as drivers or public transport passengers. A car in traffic is simply taking in the fumes from the car in front and trapping them in a box with you; the same thing happens on a bus, with a slightly larger box. Cyclists and pedestrians can to some extent choose their route to avoid the nastiest air quality, but by definition most car and bus passengers will be on the busiest routes. Underground or metro systems are even more frightening; particulates and sulphur dioxide levels can be four times as high as at ground level, with metal dust and carbon in the air as well.

That's a long description of a problem, and it doesn't even touch on greenhouse gas emissions. Nor on the obesity and mental health crises, although these are also urgent problems, and specifically problems of urban life. In 1999, the *American Journal of Preventive Medicine* coined the phrase 'obesogenic environment' to describe the way which some patterns of land use make it almost impossible to take exercise. Town planners have known for years that mental health problems are exacerbated by factors like lack of green space, noise and pollution and access to exercise. And all these problems are at least in part attributable to the single fact that cities have been designed for cars, not people.

The good thing is that there's a solution. Peter Murray has looked at success stories all around the world and worked out a sequence of steps and policy changes that can be used to promote cycling. And it's possible to be ambitious; in Amsterdam 38 per cent of all journeys are taken by bicycle (while in Groningen, Simon Koorn's home town, the figure is 58 per cent).

I don't want to oversimplify Peter's and Jan Gehl's work, which involves an entire vision of town planning. But the key to it is separate infrastructure; in order to cycle in a town, people need to feel safe. That means separating cycle traffic and taking space away from cars. In Copenhagen, the standard is to have one-way cycle lanes, which are placed on the kerb side of parked cars rather than the road side, letting the cars themselves act as a separator. If budgets are a problem, it's possible to start off with cheap temporary solutions, separating the cycle traffic with planters or benches, but the absolute key is to provide physical protection and to make the cycle network continuous. Small lanes marked out with paint or coloured tarmac don't attract the mainstream city resident; they're still dangerous, and so you're still only speaking to that proportion of the population, usually male and in good physical health, who have slightly higher risk tolerance and who are at least partly treating it as a sport. The test of cycling infrastructure ought to be whether you would allow a ten-year-old to cycle to school on their own along this route.

The distinction between sport and transport is one reason why I disagree with people who say showers in workplaces have something to do with cycling infrastructure. It's a nice gesture, and well meaning, but cycling ought to be a faster alternative to walking, not a substitute for running. If you want to go from one office to another in the middle of town, you don't change into your running gear, put on your running shoes, run a mile in ten minutes, then walk into the next building and try to find a shower and changing room to get ready for your next meeting; you just allow the correct amount of time and walk. The idea that you need to take a shower after a bicycle ride simply shows that people don't take cycling seriously as a mode of transport. Nobody feels the need in Amsterdam or Copenhagen, and it's not because they don't mind being sweaty. The bike traffic in those cities moves at a sensible pace which is set by the majority of the city residents.

The other great cliché of half-hearted 'cycle friendly' policy talk is the 'last mile solution'; the idea that cycling or walking is really only used in cities to get you to and from the train station, or some other public transport hub. Along with showers in workplaces and electrical vehicle charge points, this is one of the three signs of a local government that's realised they have a problem, but that hasn't really understood that they need to rethink things completely. Although public transport is better than private cars in that it takes up less road space per person, it's still really quite bad. The infrastructure is often incredibly expensive when you look at what could be achieved for the same money in light electrical transport, and the pollution factor is often worse; travelling on most underground rail systems is like chain-smoking metal-laced cigarettes. This sort of thing unfortunately gets politicised, because for so long the debate was focused on cars versus rail, so people assume that if you criticise one then you must support the other. But the possibilities have changed. Nearly all journeys within cities are less than six miles, much less than the range of an electric bike. And moving the population to smaller, lighter vehicles would mean that traffic problems don't need to be solved. Putting people into tunnels underneath the ground was only a good idea when it wasn't possible to have them all out on the streets where they belong.

Cycling policy is just as simple as that. It's a matter of reclaiming space in order to let people ride bicycles without worrying about dying or having their bike stolen. It's not expensive – the Dutch authorities spend about 25 euros per person per year in maintaining cycle networks and building new infrastructure. In the world of town planning that amount is regarded as a huge investment, but it's tiny compared to the hidden subsidies given to car owners. Even in terms of directly attributable cost and benefit analysis it pays for itself; the benefits in terms of wellbeing, mental health and obesity can just be left as additional gains.

Unfortunately, there's a political proverb that says: 'We know what to do; we just don't know how to get re-elected once we've done it.' Taking space away from cars is politically difficult, because car owners have had everything arranged for them for so long that they regard their subsidies and benefits as property and rights. Even the smallest compromises like 'low-traffic neighbourhood' schemes seem to trigger passionate reactions from a small but angry group of drivers. It takes a confident politician with broad electoral support to stand up to them. During the pandemic in 2020 and 2021 several London councils started to take advantage of the reduction in road use to rethink their spaces and create low-traffic neighbourhoods, but many of them subsequently dismantled the new street barriers and cycle lanes because they were being vandalised by car owners who couldn't bear to lose their shortcuts.

The Brompton is not only for commuting but also for adventure.

Interestingly, though, there appears to be a silent majority in favour of change. If you listen to populist media, you would think that traffic reduction schemes and cycle infrastructure were ridiculous ideas being imposed on the population by politicians. When people do rigorous market research and polling, though, they find that these things are popular. It's worth remembering that, as I mentioned earlier, half of all households in London don't own a car at all, while in New York that figure is more like 55 per cent. In Tokyo, the city authorities won't let you buy a car unless you can show that you have somewhere off the street to park it: as a result, nearly 60 per cent of households don't. Car ownership is a minority interest in most cities of the world; it's only in places where it's more or less impossible to get about in any other way, such as Houston, that everyone drives. Even in Los Angeles, about an eighth of households don't have access to a vehicle, most of them living in the city centre. And many car owners in cities don't enjoy the way the roads are organised. A lot of the most vocal members of the motorist lobby are actually campaigning for their right to live on the outskirts of town, then drive in and park their car as if nothing had happened to the traffic since 1977.

It ought to be possible for politicians to recognise that although they could gain short-term political advantage by continuing to provide car owners with their public subsidies, doing so would be disastrous in the longer term. Whatever their party, they must be aware that in twenty years' time either they will be out of power or the modern equivalent of the horse manure problem will be on their desk.

This is one of the most frustrating things about the political failure to make the necessary changes; the economic and demographic forces are basically impossible to resist. Everywhere is on the same curve. It may be possible to delay the inevitable a bit longer if you have lots of spare space, or if things are slowed down where the car lobby is particularly strong or the political culture is particularly weak and opportunistic. But eventually the same thing will happen as happened in Amsterdam; the city will reach a point at which the dominance of the car becomes intolerable. Amsterdam reached this tipping point earlier than almost any city in the world; it has a medieval grid which wasn't much affected by the Second World War, and by the end of the 1960s canals were, insanely, being filled in for conversion into roads. Often when people look at what happened in Amsterdam, they identify the oil crises of 1973–4 as the catalyst for change, but in fact it seems to have started earlier than that, and as a reaction to the way that car infrastructure was affecting people. The first serious campaign group to start publicly advocating anti-car measures was a women's organisation in the 1970s called Stop de Kindermoord ('stop the child murder'). Groups like this are growing up all across the world today, and it's poignantly noticeable that many of the cycling advocacy groups are founded and run by people who were radicalised into action by friends being killed.

Political leaders can either speed the process up at little real cost, or slow it down at the expense of lost health and human unhappiness. Anne Hidalgo, mayor of Paris since 2014, has committed to putting in 650 kilometres of segregated cycleways and taking out 80,000 parking spaces, for example. Despite initial protests, she was re-elected in early 2022 for a second term, suggesting that this kind of change isn't as unpopular as politicians fear. Cities like Shanghai have gone from ubiquitous cycling to being choked with automobile traffic, and back again to government restriction of car ownership by a system of permits, all in the time that London has taken to introduce a congestion charge, extend its area once and then change its mind about the extension. It's depressing that more cities don't make the right choice.

All the way through this book, I've been saying that making plans and finding solutions has to start from a position of respect for the problem. Cities have to change, but the way in which they are likely to change is under the control of politicians who might not make the right decisions or make them fast enough. Brompton isn't in charge; we're only one part of the system, so we need to react to the conditions as we see them and make sure that we're providing resources when they're needed without causing more problems ourselves.

Does this mean 'make more bikes'? Yes. There is no better solution than portable, human-powered transport, with electric assistance to make it comfortable for the user. It's cost-effective, it's not obesogenic and it's durable rather than disposable – some of the kit used by urban electric scooter startups seems to have a terrible rate of depreciation, but there are ten- and twenty-year-old Bromptons still being sold online for close to the normal retail price. If you look at 1920s photographs of city centres, particularly in Asia, you can see a model that would work today. These photos belong to a past age where people wore the same clothes for longer than one season; part of the problem today is that a lot of companies then decided, based on an incorrect assessment of the planet's resources, that it was good for business to design things that would keep needing to be replaced. Consumers became used to looking at price alone without considering durability or ongoing usefulness, an approach that people have only recently begun to question.

In order for Brompton to make a contribution to solving the problems of cities that is of the same order of magnitude as the problems themselves, we would need to scale up a lot more – probably by a factor of ten at least. And that sort of scale of change would probably once more mean that we couldn't keep doing things in the same way. If we adapt our old motivational slogan and ask 'What would this look like when we are making 500,000 bikes a year?', it takes us to some very unfamiliar places.

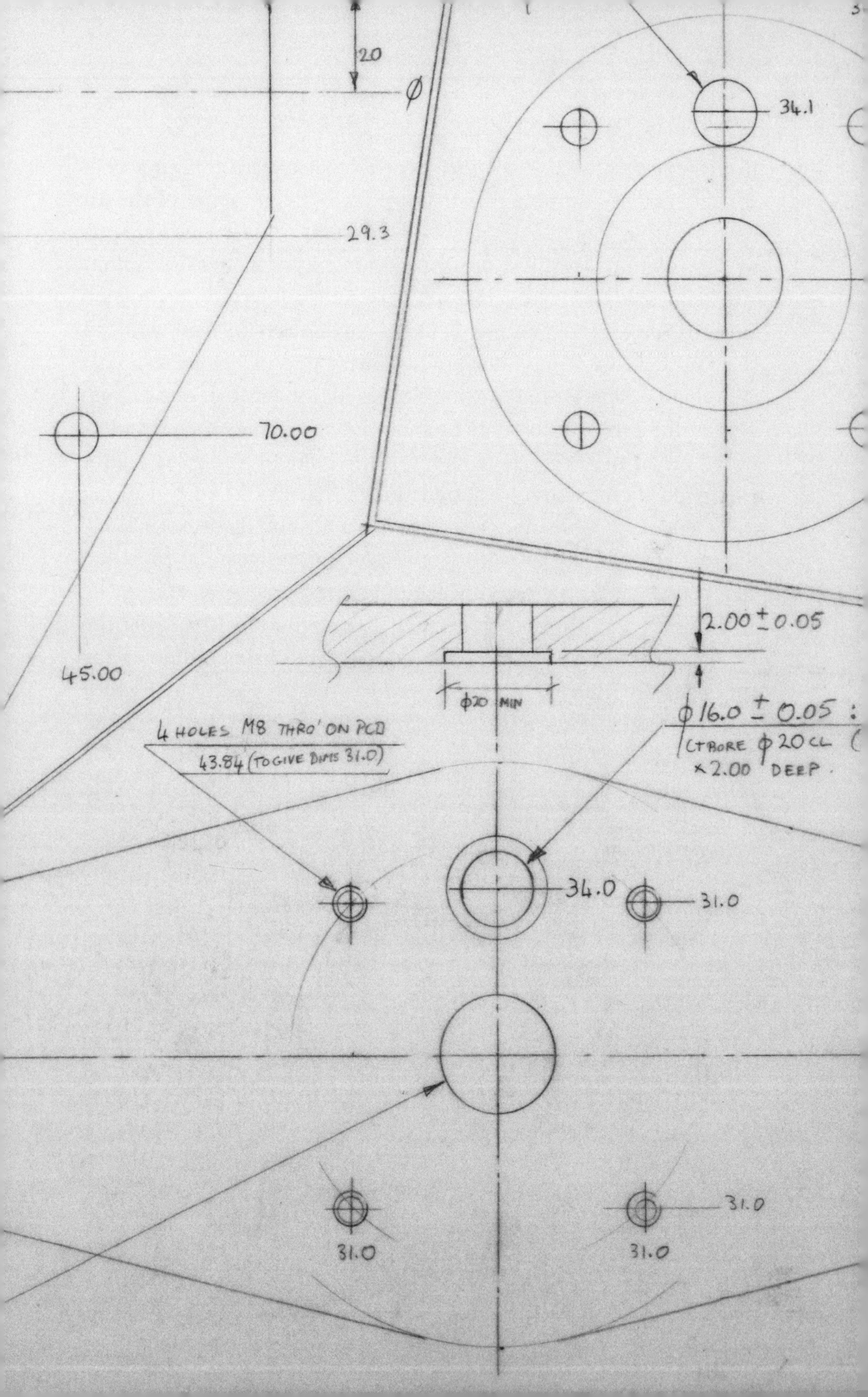
20
29.3
34.1
70.00
45.00
2.00 ± 0.05
ϕ20 MIN
ϕ16.0 ± 0.05 :
CTBORE ϕ20 CL
× 2.00 DEEP.
4 HOLES M8 THRO' ON PCD
43.84 (TO GIVE DIMS 31.0)
34.0
31.0
31.0
31.0
31.0

13 INNOVATING FOR THE FUTURE

The electric Brompton took more than ten years to develop, from the initial proposal to the first models arriving in shops, but the original Brompton only went into mass production fifteen years after Andrew Ritchie drew up his original design.

Tony Castles is an electronic engineer who suffered a serious spinal cord injury in a bicycle accident in the 1990s. As he recuperated, he found that cycling was sometimes easier than walking, but that going uphill was difficult. What he wanted was a bike that he could carry and put in the boot of a car, but that would also be easier to ride up hills. And what he came up with was a system that used a 4 kg battery and a clever motorised hub gear for a conversion kit to turn a Brompton into an electric bike.

We were impressed. There is something about the Brompton which encourages people to modify it: we have loyal fans all over the world, and many are engineers who want to adapt the design to suit their exact requirements. People try to build recumbent Bromptons, three-wheelers or cargo trailers, as well as lots of replacement components to cut weight or adjust the riding style. We generally tend not to encourage them. Even though these are some of our most dedicated fans, and they often come up with ideas that look fantastic, they don't always fully understand the design and lifetime loading of the bike. Most of even the best ideas in the 'pimp my Brompton' community are just a bit too dangerous for us.

The engineering of the Brompton is a lot closer to the limits of tolerance than it is for most bikes. This means there is a lot less room to make changes that push things outside the envelope of our own testing. A chain pusher plate machined out of titanium to save ten grams of weight probably won't have much effect on the stresses placed on the frame, but many other modifications might. For example, there are plenty of merchants on Alibaba and similar websites offering carbon-fibre Brompton frame parts. When we started using carbon-fibre parts ourselves, it was after a long and rigorous round of finite element analysis, stress testing and many design iterations. We certainly can't certify the safety of a carbon-fibre manufacturer who has never even given us a phone call.

For this reason, we have a blanket policy; if you modify the bike, then you have invalidated your warranty. It seems harsh, but it has to be clear. The only way that we could continue to put our warranty and name behind a modified bike would be for someone at the factory to put in a hundred hours driving themselves crazy on a fatigue rig in order to satisfy ourselves that the modification didn't affect the safety or lifespan of the bike as a whole. And it's obvious that we couldn't possibly commit ourselves to doing that; the pimp-my-Brommie community is just too big and too inventive to keep up with.

Nevertheless, we've always taken an interest in really inventive ideas. From a legal liability point of view, we couldn't endorse Tony Castles's project, but nor could we resist seeing how he got on. I was intrigued. We were able to sell him some parts which he needed, and I corresponded with him while he brought it to market as the Nano conversion kit in 2006. This was when Brompton started to realise that, as a company, we were likely to be defined over the next couple of decades by the approach we decided to take to the electrification of bicycles.

Electric bikes were one of those things like the internet – you either immediately saw the point or you didn't. If you saw it, you just knew that this was the future and that the whole industry would need to adapt to it. If you didn't, they looked like a slightly embarrassing fad. Now, with the benefit of hindsight (and again just like the internet), it's almost impossible to believe that anyone could have been unconvinced. And indeed many people have found it useful to make a few mental adjustments to the historical record, in order to put themselves closer to the winning side.

Hand on heart, I can honestly say that I was one of those who got this right very early on. The first time I rode a bike with electric power, I knew there would have to be an electric Brompton. It reminded me a bit of the Solex, the French petrol engine 'assist motor' bike that you can see in Jacques Tati films from the 1950s. When I was a kid, I once spent a holiday in a rented house in Brittany where my cousin and I found some of these old motors in the garage. We spent three days getting them to work again before riding them around like young kings. On the fifth day we were caught by the gendarmes, because in the period since the company stopped trading the Solex had been made illegal to use without a motorbike helmet and licence.

A similar regulation has caught up with the early electric bikes. If all you're doing is twisting a throttle to apply electric power to the axle, then that feels more like powered transport. In many countries including the UK it has now been ruled that you have to be making some contribution with the pedals if you are going faster than 4 mph and want still to be treated as a cyclist. In any case, the modern 'pedelectric' system is more pleasant to ride. The way it works is that there's a permanent magnet in the hub, which is surrounded by electromagnets. The electromagnets

Will Carleysmith setting off at pace at the 2007 Barcelona Brompton World Championship.

are controlled by a microprocessor, which effectively switches them on and off so that there's always a magnetic attraction pulling the wheel in the same direction that you're pedalling. If the timing and current going to those electromagnets is just right, they provide the same sort of subtle addition of force that gravity gives you when you're going down a gentle slope, or when you have the wind at your back. In order to get that seamless feeling, though, the bike needs speed and torque sensors to detect what you're doing with the pedals, and some quite sophisticated software to change the pattern of switching on the motor in real time.

When it works, the feeling of fun is intense. But more than that, it was perfect for Brompton. The whole point of the company is that our bikes are meant to be a mode of transport, not for sport. Having a folding bike means that if for whatever reason you don't feel like using it, you can jump on a bus instead. Being able to push on the pedals and get a power assist is just another option to keep you moving. It also helps you to commute, because you don't work up a sweat – the reason why people need a shower after cycling to work is because they've been trying to go too fast, or because they've had to make an effort going up hills or accelerating away from lights.

A lot of resistance to adopting electrics in the early days came from sporting types who thought it was cheating to ride a bike without doing exercise. I didn't think there would be much of that kind of opposition at Brompton, and there wasn't. There are always some cycling purists, but the basic logic of having an electric Brompton was fairly unassailable. But people within the company did have two other substantial objections to electric power: weight and money.

In order to get an electric bicycle to market without compromising our principles, we would need to get the overall weight – now increased by motor plus battery – down and back into the range of reasonable portability. This would require considerable technological and engineering innovation some way beyond the state of the art in 2006, because the first electric bikes were huge. And we would also have to find a way to do it without spending so much money that it put the company at risk. And so began an extraordinarily agonising ten-year process, with multiple false dawns, arguments, embarrassments and at least one threatened resignation on my part. Tony Castles couldn't have known what he was starting.

We began the project by talking to Tongxin, the Chinese manufacturer whose motors Tony had been using. The motor they made was pretty nice, weighing in at just over 2 kg. It had an unusual gearing system which used friction rollers rather than actual gears. This was a clever idea; it kept the motor quiet and allowed it to be a bit smaller. As long as you don't try to transmit too much power, friction gears can be pretty efficient; you are basically trading off the losses to friction in the teeth of conventional gear wheels against the losses to slippage between the rollers. Unfortunately, it seemed to us that the trade-off was not going to last. As the motor aged, the rollers would gradually polish each other to a smooth finish until the whole motor would need replacing, far short of the life that we wanted for the bike,.

We thought this design issue could be addressed, however. So we got in touch with Tongxin, and went to see them. They met me at offices in Hangzhou, about two hours' drive from Shanghai. The offices were perfectly pleasant city-centre premises, but nothing was being made there. This was disappointing – it's a feature of modern life that we seem to hide away the business of making things as if we were ashamed of it. I got on well with the people, though, and we even went for a cycle ride together in the park so that they could see the Brompton in action. They weren't prepared to take me to see the factory, but they seemed enthusiastic about the idea of working with us.

On balance, we were optimistic enough to go back home and talk to the designers. Will Carleysmith had joined Brompton as a graduate and worked his way up to become the chief designer, taking the desk and position once occupied by Andrew himself. When we had decided on what we thought might work as a modification to the Tongxin design, it was his turn to go out to China, this time with strict instructions to insist on seeing where the motors were made. Will was driven for several hours into the countryside, to a sort of industrial estate where the main industry was a very large pig farm. Even with the best management in the world, a large pig farm is going to smell bad, and this one certainly did. In a nearby shed, Will was shown Tongxin's production line. Rows and rows of workers were standing at benches, winding the coils for the electric motors by hand, slopping on glue for the magnets, and all the while smoking clay pipes.

This frightened us. It was the immediate end of this avenue of exploration. There was no possibility of designing quality into that process, and we could hardly inspect the defects out from the other side of the world. We hadn't even started to look at the batteries, which could have had all sorts of risks and issues of their own, but since we couldn't trust the manufacturing process and there was

The Brompton needs to be light, as not all riders are physically strong six-footers.

no prospect of changing it, there was no prospect of continuing. It was a very clear illustration of why we have the 'invalidate your warranty' policy.

At this point we realised that nobody we wanted in the electric motor industry seemed to want us. Big manufacturers like Bosch or Heinzmann had great quality and made reliable motors, but they weren't interested in the size of order that we could realistically place. They either gave us a straight no, or a qualified yes, but with so few changes possible to their off-the-shelf product that the weight and size made it pointless. At times we toyed with the idea of taking one of these motors and putting it onto a Brompton anyway, just so that we were in the game, but only bodybuilders or rugby players would have been able to carry one.

It ought to be obvious, but it's actually a hard-learned lesson; the trouble with innovation is, the things you want to buy don't exist yet. There seemed to be no middle ground between the big producers of big runs of big motors, and the small, innovative companies whose processes didn't stand up to scrutiny. But then we found Ultra Motor. They were one of those venture-capital-funded companies who were aggressively chasing the dream of being the Ford or Microsoft of the global electric bike market. They had some underlying technology invented by a former Russian radar scientist, an R&D team originally from Germany but now in their factory in Taiwan, and a host of joint ventures trying to bring e-bikes to the largest possible 'total addressable market' in places like India and Vietnam. This sort of world-domination outlook meant they were keen to work with us too.

At this point I broke whatever law it is that tells you to say either what you're going to deliver, or when you're going to deliver, but not both. We had met Ultra's designers, seen their factory and signed contracts, and were encouraged by their European sales director who was a congenital optimist and fantastic at his job. So we announced to the market that the electric Brompton was on its way, and told them when it would be made available.

There was a hard lesson to learn here too; the publicity benefit of big splashy launches comes with a risk attached. If you promise something big and don't deliver, you damage your credibility. The technological challenges soon turned out to be much harder to overcome than anyone had anticipated at the planning stages. Making things smaller often creates new problems – the materials may not be strong enough at the reduced size, or the tolerance may not scale down along with the size of a component. And in an electric motor, the closer together things are, the more likely it is that they will have electromagnetic interactions with each other, or that heat will build up where it's not wanted.

Ultra also seemed to have some of the typical problems with focus which often affect venture-capital-backed companies. Halfway through the project, its UK bike manufacturing subsidiary stopped trading; then the manufacturing and design divisions were sold to a different group of investors. The things that they had promised to deliver just didn't arrive. It became increasingly clear to us that we were not going to get a working electric Brompton through this route. Having to apologise to the whole bicycle trade and wasting so much time were if anything more painful than the money we lost, which itself was nearly 10 per cent of the year's profits.

The mood in the small team working on the e-bike project at Brompton was understandably low as we tried to recover from the humiliation and start our third attempt. Another thing we had learned was that we needed to work with a manufacturer that was entirely based in Europe. Working with Ultra had made it clear that a project for something new needs a lot of meetings and communication. You're effectively building a new organisation from scratch. Different elements of the new product require particular, specialised knowledge, but because it's a new product the different components may never have been put together in that way before, so the people with the understanding of their own parts don't necessarily know how they interact with the rest. This knowledge has to be established by experiment and trial and error.

The moment you start putting things together and machining parts, even for the roughest prototype, you seem to turn up new issues and questions which never occurred at the design stage. At that point, everyone needs to get together and decide what has to be changed in order to solve the problem, and how that change will affect other parts of the design. If every iteration of this process requires you to co-ordinate calendars and get a flight to Taiwan, it takes forever unless you are extraordinarily lucky. Even with remote working and online communications, a wide time-zone gap radically shortens the usable working day, which matters a lot as people often need answers to important questions outside those hours.

We stumbled down a few more blind alleys in the search to find our 'unicorn': a not-too-big, not-too-small electric motor company, teeming with innovation, interested in the concept, willing to work with us, and located somewhere near enough to west London. A manufacturer of electric wheelchairs was interested in principle but unwilling to commit, and this reflected the stage that the market had reached – by now, there was so much demand for e-bikes that suppliers were reluctant to divert capacity and effort from anything other than satisfying the immediate demand. I could sense things slipping away.

In the end, the solution came from a friend. In my experience, if you make enough connections and give yourself enough opportunities, sooner or later one of them will end up coming good. Patrick (now Sir Patrick) Head was the co-founder and head engineer of the Williams Formula 1 team. He had a proper motorsport lifestyle – when I called him in 2013 asking for help he was on his yacht in the Mediterranean – and was an enthusiastic user of his Brompton to get him from the Battersea helipad to his meetings in London. That is not necessarily the median-use case for urban mobility, but it still counts. I hoped that he might know of an obscure electric motor manufacturer that we hadn't tried yet.

It turned out that this was exactly the right time to call. For a few years, Formula 1 teams had been using 'kinetic energy recovery systems', or KERS, to get an advantage at corners. This was similar to the way that a hybrid car works – slowing the car down by making the wheels drive a dynamo instead of by braking them, and storing the electricity generated in a battery. The main difference between a KERS car and a hybrid car is that the hybrid will sometimes use the electric motor as an alternative to the petrol engine, while the racing car always needs to use them both at the same time – the stored energy gives an acceleration boost coming out of the corner, using the high-school physics-lesson principle that the same arrangement of magnets and coils can be a dynamo or a motor depending on whether you're putting energy into it or taking energy out of it.

What this meant was that Williams, over the past four years, had learned a lot about making very light and efficient electric motors, very light and efficient batteries, and very quick and reliable software to integrate the boost from the motor with another power supply. If I had called Patrick two years earlier, their KERS system would have been a trade secret, but a couple of years before, in 2011, they had launched Williams Advanced Engineering, a new subsidiary with the mission of finding new commercial applications for all the things they had invented. In principle, our problem was exactly the sort of thing they were looking for. The electric Brompton wouldn't need kinetic energy recovery, it just needed a small and light electric power train that could work with the pedals. Patrick introduced me to Craig Wilson, who was leading the Advanced Engineering venture, and I felt like a drowning man who had been thrown a rope. It couldn't be long now; surely we were going to catch up with the market and make good on our promises? In fact it would be five more long years before the first electric Bromptons went on sale.

Part of the problem was money. Car-racing engineers don't come cheap, and Craig was refreshingly candid up front about the fact that this project was going to cost a lot. A government grant called 'InnovateUK' made all the difference here – we were awarded £200,000 from a scheme for encouraging and funding advanced research and development in industry. This award meant that

The Brompton dock, with a sleek design and stand-alone solar power.

Brompton were just about able to fund the project without risking the company's financial stability. It also helped a great deal in discussing things with the board and getting their agreement to invest our own capital; coincidentally, it was also nearly the amount which we had lost on the Ultra episode. Furthermore, the grant was awarded as part of a competitive process, judged by technical and business experts, so having their seal of approval as disinterested third parties went a long way towards convincing waverers on the board. Williams were also willing to take the last £100,000 or so of their fee in the form of a royalty on the first 10,000 electric Bromptons to be sold. It was one of those deals that makes you slightly giddy, like sending a wire transfer off into the Russian banking system for titanium six years earlier.

The challenges began shortly after we started work. This was quite an important learning experience for me, and the lesson here was that when you have problems in a situation like this, you can't presume that it is because there is something wrong with your partner company. If a motor supplier is based in a pig farm, or your castings company hasn't swept the floor since the 1920s, you can say that it's all their fault and if you were working with better people there would be no problems. When the partner is an impressive operation like Williams (or, looking back on it, Haas), you don't have that to hide behind.

On the other hand, the team at Brompton working on the electric project were no slouches; we started with just myself and Will Carleysmith, then hired David Rhys to be head of electric systems and built a team of five exceptional engineers. Nobody involved in the project at our end was anything less than very capable, which meant that the problems had to be intrinsic to the process rather than attributable to anyone's personal or professional failings. And the main source of problems was simply that Brompton and Williams were two separate organisations.

When you sign a partnership deal and start working with another organisation, there's a tendency to draw two boxes in your mind or on a whiteboard, and add a double-headed arrow to indicate that they're connected. In fact what's happening is that a dozen of your individual people are working with each other and with a dozen of their individual people, and the number of arrows needed is well over a hundred. This explosion of combinations is part of the subject of a good book about software engineering by Fred Brooks called *The Mythical Man-Month*, and it's his reason for believing that adding more manpower to a late project will just delay it more, because the amount of communication overhead added is greater than the marginal contribution of the new engineers.

The networks can be made to work as long as they are stable. But people move, change jobs and get assigned to different projects. Within an organisation this is disruptive enough, but at least you get warning of it happening and you're able to arrange a handover. When two organisations start having to deal with one another, unless a huge amount of effort is expended on making the communication happen and preserving all the accumulated knowledge to date, every personnel change turns into a major setback. Very few organisations can handle this well.

But after three years, the project was finished; the Williams team had worked miracles and met the specifications of an engine and battery that could be fitted to a Brompton and give us a fully functioning electric folding bicycle weighing 15.5 kilograms. We paid the money, took a deep breath and then took stock of how much more of a mountain we had to climb. Of course, a Formula 1 advanced engineering company isn't going to deliver a product that you can just start manufacturing immediately. The whole basis of their industry is that they make highly customised small batches, and work on the very edge of the performance trade-offs. (There's a proverb variously attributed to Enzo Ferrari, Colin Chapman and other industry legends to the effect that the ideal racing car would be one that crossed the line a metre ahead of its nearest rival and then fell to pieces.) We needed to take the motor and battery system that they'd developed and find a way to produce it in the thousands, with the same reliability for the customer as the pedal Brompton.

This is the nature of research and development as it happens in companies. There are not many firms in the world doing cutting-edge science; in all probability, the solution to your problem has already been found and is already out there. So you have the problem of knowing what's going on in your industry and finding the right tools and partners to bring together in the same place. Those elements might be anywhere in the world – they might be in the local university, in a factory on the other side of the world doing something completely different, or in your own plant on the desk of a paint inspector who you never knew was able to programme a control system. This kind of research happens by serendipity to a substantial degree, but it's not pure luck; it happens faster and better the more that a company is set up to be open to the world and to take ideas seriously.

Finding a solution to the problem in principle and building a prototype lets you then move on to trying to develop an actual product. Size and scale help a lot with this stage of innovation, because they let you put resources into something that's not immediately generating revenues. One of the big benefits of profitability is that you have more control over what you're doing, and this allows you to take the time and effort to get something right and make it reliable. Sadly, there's no short-cut past this in an industry where you're putting your product onto streets where people can get hurt. A large part of the business of R&D is testing, over and over again, to make sure that the idea works.

But before any of those things can happen, the original idea has to be there: you have to invent the problem before you can solve it. Some ideas are obvious from the nature of the product – we want the bike to be light, and that has driven solutions from simple things like 'take the gears off' to the cutting edge of titanium engineering. Some are a matter of spotting trends, like the benefit of electrics. In many cases, the idea comes from listening to the customers; another advantage of scale. But the moment has to be there; perspiration (and organisation) is how things are brought to market, but someone has to have the big dreams like 'let's go to the moon' or 'what if you could carry an electric bike in one hand'.

By 2017, the electric Brompton project had to start showing some revenue or it was going to die. This really matters in any company which has a board; you'll recall that I had specifically got their agreement to invest capital in electrics, and I'd spent a lot of time persuading them that it would progress a lot better than it actually did. For as long as a project is a pure cash drain, it's at risk of losing the confidence of board members; unless people believe in the story, they're not going to keep funding it.

Once the project has some revenue attached, however small, it's a lot harder to disbelieve in the story. People can now see that there is a product which someone is prepared to pay money for – it's no longer pure speculation. Better still is when a project reaches break-even: when that happens, it stops being an agenda item and it will continue unless someone specifically wants to kill it. But the first revenue is a crucial milestone which makes it so much easier to justify more resources. The problem was that the team were still saying the electric Brompton was not ready.

We were approaching the point of no return. If we didn't want the project to fold, there was only one way to go: we had to take the plunge and say that it *would* be ready. We decided to set a date, in the full and painful knowledge that plenty of people in the bicycle business still remembered the last time we had done so, and this time stick to it. But rather than repeat the previous mistakes, I was going to take a lesson from Andrew and the first-ever Brompton bikes.

Andrew always said that the five-year gap between making the first 500 Bromptons and being able to restart production saved the company; the feedback from those early users allowed him to change the design and solve a fatigue problem that would have been disastrous at scale. So we planned to do something similar with the electric Brompton. We let the news slip out gradually, through specialist press and user clubs; channels where we could be reasonably sure that existing Brompton fans would find out first. And we restricted the first manufacturing run to just 500 bikes. When we sold those first 500, we made a generous allocation to warranty costs, and made sure that the buyers understood that they were our pioneers, we were prepared for teething troubles and we would look after them.

It seemed to work. Things weren't perfect, but we'd planned for that. We improved the bearings, changed the seals, rewrote the software that controlled the delivery of power and made a number of improvements to the manufacturing process. We got so much feedback that we ended up inviting all 500 owners to a party to thank them for it, and even now, years after the warranties have formally expired, we will look after any problem on those first pioneering bikes. Over the course of another couple of years, we reached the point at which we felt able to start really marketing the electric.

From my initial conversations with Andrew about Tony Castles's conversion kit to the stage where Brompton had a working electric model took ten years – not really all that much less than the fifteen years from Andrew's first designs to the beginnings of mass production of the original Brompton bike, particularly when you allow for the hiatus when nobody would lend him money to start production. It's almost as if there's a balance between opposing forces; the greater the resources that are available to you in terms of testing equipment, computer analysis and rapid prototyping, the bigger the tasks you take on, and so the overall time to get something to market is close to a constant.

That ten-year journey to bring a new idea to fruition also taught me something that wasn't obvious to me at the start. A large part of the reason for things taking more time and costing more money than you expect is that people and organisations are different, and those differences manifest themselves in communication gaps. Getting information across those gaps in both directions is something that doesn't just happen by itself, and making it happen is a huge part of management.

Beyond that observation, I'm wary of trying to draw too many conclusions from the experience, because I don't think that innovation works like that. You can see from the story of the electric bike that there were plenty of false moves and avoidable errors: things that were obvious in retrospect. It would be tempting to do a lessons-learned exercise and draw up a process to make sure that the same mistakes aren't made in future projects. Except that future projects won't have the same mistakes to make, because they will be different. The whole point of doing something new is that you don't know what's right or wrong. Prematurely identifying something as a 'mistake' because it looks a bit like something that happened on a different project in the past is going to ensure that you don't do anything genuinely new. It's probably even wrong to call the blind alleys of the e-bike project mistakes – they were part of the process of learning, and it's a delusion to think that the learning could have happened without the process.

What you can do instead is concentrate on the consequences of making mistakes. Any rule which boils down to the empty platitude of 'don't make mistakes' is futile; a better rule would be that as long as the worst case isn't too bad, you might as well try it. I've always allocated a sum of money at the start of every year for people in the company to have ideas with. One day this may get institutionalised and then we'll have a fund with a grand name like Greenford Innovations or Brompton Future or something, but at the moment it's personally under my control and it's called the 'Fuck-It Fund', because that's what I generally say when giving someone permission to try something. The fund has grown over the years, but the principle is always that if someone has an idea that they're passionate enough about to approach the chief executive with, and it won't damage the brand or cost an absurd amount of money if it all goes wrong, then they can do it.

And when it comes to new ideas, we don't do corporate planning, discounted cash-flow forecasts and returns on investment. I did a few of those at ICI when I was an engineering project manager, making investment cases to go up to the committees that decided on new equipment or facilities for the Melinar plant. The experience taught me that if someone wants a project to go ahead, they can fudge a spreadsheet to deliver the required returns; conversely if they want to kill a project they can fudge numbers the other way. Since the returns calculation is always more or less dependent on how much someone wants to do the thing, you might as well cut out the middle stage. Not only do you save time and effort, but you avoid the issue of accidentally running a training course for your employees in how to lie to you.

So for the purposes of the Fuck-It Fund, the return-on-investment calculation is simply 'what would be the worst thing that could happen?' If the answer is just 'we lose all the money we spent', then as long as it's not a silly sum, you might as well do it. If the idea works, then we've got an unexpected benefit. If it doesn't, then that's what happens to most ideas, and since we write off the money put into the fund at the start of every year, it doesn't affect the financial results when we spend it. And anyway it's rare to find a totally failed project. By trying different things and making a mistake, you're actually creating real value because you have tried it on a small scale and found that it didn't work. It would be much more dangerous to justify it with a business case, make a big investment and only then find out that it was a mistake. Even an otherwise wasted trip to a foreign trade show can be a better team-bonding exercise than an afternoon's paintball.

In other words: try things if they aren't disastrous, and stick at them, because finding out information takes time and disappointment. I try always to make sure that someone else takes over my big ideas, partly because I know that I prefer having big ideas to seeing them through, and partly because any project that is going to turn into a real business that employs people and consumes resources needs to be owned and taken care of by someone who can dedicate themselves to it. So the titanium operation in Sheffield, our joint venture in China, the e-bike, even the new factory – these were all my ideas, but I've given them away; they're somebody else's sweetheart now.

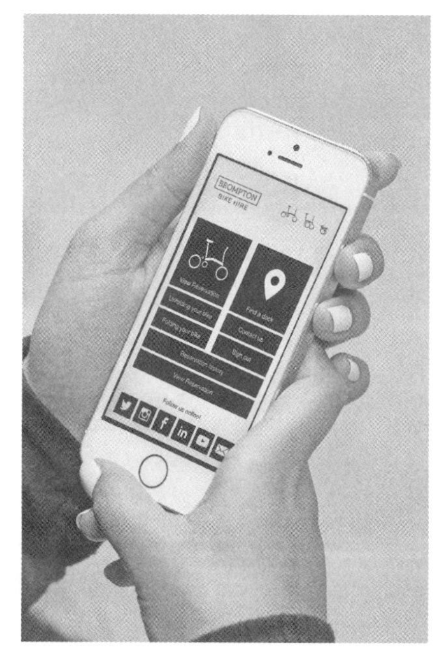

Reserve a bike to pick up, and drop off at any dock in the network, with the Brompton Bike Hire app.

One of the ideas which remains my personal responsibility, though, is the Brompton Bike Hire scheme. And it's still identified with me because it hasn't worked. Yet. We first launched this scheme in 2011, and it's gone through two or three board meeting votes over whether to give up and cancel it. So far it's survived, and in 2021 it came close enough to breaking even as to make no odds. But along the way, some of the annual losses have been big enough to be fairly embarrassing to me – in a year when the whole company makes £300,000 profit and the bike hire subsidiary has losses of £200,000, people are going to ask tough questions.

I'm still personally convinced that the idea is a good one. A lot of the smartest and richest people in the world seem to agree with me, because in the eleven years that Brompton Bike Hire has existed, venture capitalists have put literally hundreds of millions of dollars into bicycle and scooter hire startups. Meanwhile, we've tried to roll out our version with as much control as possible over the financial exposure; as with everything else we've tried to do, the overriding consideration is that, in the worst case where the whole plan falls apart, the existence of the company isn't jeopardised.

The basic unit of Brompton Bike Hire is a 'dock' – effectively a box roughly the size of a medium-sized caravan, which is divided up into forty lockers, each of which contains a Brompton. The lockers are connected to the app and central booking system through which people take the bikes out and rent them. It's not a difficult concept; the only significant difference from other bike rental schemes is that the bikes are in lockers, shielded from the elements and from vandals. That's necessary, of course, because they're more valuable and easier to carry away than normal hire-scheme bikes, and it's practical because they're small and foldable. In business terms, we sell the docks and the bikes within them to the owners of their locations, as an additional amenity to their buildings and railway stations, then we take the ongoing hire fees in return for servicing and maintaining the docks and bikes. The first one was installed in 2011 in Guildford, and there are now fifty-one of them around the country.

This lets us finance the establishment of the network, which is the expensive bit for all venture-funded startups. It also spreads out the risk, because the Brompton Bike Hire network is made up of lots of individual little deals with different site owners, but the docks are entirely interoperable with each other. Many of the cycle and scooter networks that have grown up have been vulnerable to the fact that they signed a mega-deal to put bikes and docking stations (which they paid for) all over an entire city, so that when the authority decides it doesn't want them any more, the entire business is gone. But because the Brompton docks all link together, you can pick up a bike at the railway station in Manchester, take it on the train down to Euston, ride it around London for a few days and then drop it off at King's Cross when you take your next train somewhere else.

In principle, it's hard to see how this idea could go wrong for Brompton; the up-front costs are largely covered and there are several different sources of revenue – it even provides a load of free advertising and a try-before-you-buy opportunity for the bike itself. In practice, as soon as you start putting anything into a public space, you begin to realise what utterly chaotic places cities are and how things can happen that you never even dreamed about. You need to proactively maintain the bikes, and return them to their owners if they've been dropped off in another dock. You need to provide 24/7 customer service, because you never know when a group of students is going to hire eight bikes for a pub crawl and then find at the end of the evening that they can't work out how to fold them. You find out that if an unprecedented mass of cold air suddenly descends from the east, the circuit boards will fail at temperatures below −12°C, and the batteries which power the whole dock will run out because the solar panels have been covered with snow for five days.

And of course if you're putting a metal box containing nearly forty grand's worth of highly desirable consumer products out on the street, you start to learn about crime. People took angle grinders to the lockers to try to get them open, among other inventive ideas. We lost comparatively few actual bikes, but often there was enough damage to the docks to require an expensive replacement, or in some cases an even more expensive redesign and redeployment. By carelessness or bad luck, sometimes the customers get the bikes lost or stolen too, so we needed to have the right kind of insurance, and so on.

I am confident that we will keep supporting our bike hire scheme across the UK because it has the potential to develop into something special. One answer to the big question of who Brompton's next customers might be could be that transport in the future might not be a matter of choosing between private ownership of vehicles and mass transit. That wasn't how things worked in the age of horses, after all. People who had a field and stables might keep their own horse, but city dwellers would rent one from a livery stable as and when they needed one. If you were coming into London from the countryside, you might ride your own horse there, but you would then get it stabled and take hackney carriages or walk to get around town.

This sort of 'transport as a service' could be the future for cities, because it's a model which recognises that cities are different from non-cities; they have different needs. Within a city, the problems of transport are to do with air pollution, efficient use of space and the difficulty of keeping healthy and exercising. That suggests a mixture of active transport and public transport, with a heavy weighting to the former. Outside cities, the problems are totally different; the distances are often long enough to make even e-bikes impractical, and the organisation of public transport in areas with low population density is intrinsically challenging.

I may be right or wrong; it's reasonable that the board should keep challenging me on this one, and it will be a lot easier to justify investing in bike hire when it starts to make a reasonable return on the investment already put in. But even if it fails, this will just be a bit embarrassing to me personally rather than something which threatens the ability of the company to control its own destiny. And for that reason, all our ideas need to be looked at not as atoms on their own, but as part of an overall process of innovation. It's the system that they're part of which is important.

CHPT3
SCHWALBE

14 THE BRAND, THE COMMUNITY AND THE COMPANY

The Brompton logo was designed by Andrew but not registered as a trademark until 2008. The typeface is similar to, but not quite identical with, a font called Gambero Regular.

In 1962, at the age of 22, Heinz Stücke decided to leave behind a boring job and an authoritarian family in Hövelhof, West Germany, and see some of the world. And so he took a bit of luggage and a bicycle and started to ride. Over the course of the next forty years, he travelled to 196 different countries and covered over 648,000 kilometres. He was beaten up in Iran, held hostage in Zambia and had his bike stolen six times (once in Portsmouth). But for the most part, he always said that he found the human race to be pretty decent.

There was no reality TV show about him, no world record attempt – he was a pure traveller. He supported himself by selling some of the startling and wonderful photos he took through a picture agency, and with an ever-changing printed pamphlet that people could buy from him for an amount of money that he usually set at somewhere near the price of a cup of coffee in whatever country he had reached. And he had a few sponsors in the bicycle industry; mainly people who were fascinated by the idea that a bicycle could set you so completely free.

One of those sponsors was Lee San, owner of the Flying Ball bicycle shop in Hong Kong. Flying Ball was one of the first and

There is a prize for the best-dressed contestant at the Brompton World Championship. And it is important not to wear Lycra!

best Brompton retailers in Asia outside Japan, and in 2010 Mr Lee introduced me to Heinz. And so it was that Heinz covered the last 10 per cent of his lifetime journey, at the age of 70, on a Brompton. He had converted to using folding bikes a few years previously because airline luggage allowances were getting meaner, but he was a bit sceptical to begin with. He didn't think that the small wheels would be able to cope with the kinds of roads that he went down or carry enough of his luggage – when you are cycling across deserts, you need to be able to pack plenty of water. But we asked him to give it a try on a short trip across the USA and he came back pretty impressed – changing the tyres wasn't so easy, but he was surprised by how well the suspension compensated for the smaller wheels in terms of ride quality. So now, if anyone asks whether the Brompton can handle cobblestones or potholes, all we have to do is point to a guy who took it 60,000 kilometres across broken asphalt, dirt tracks and desert highways.

You don't use the phrases 'brand ambassador' or 'influencer' when you're talking about a man like Heinz Stücke – it seems almost blasphemous to do so. But it was inspiring to have been a small part of his adventure. We ended up working with him on a book compiling some of his photographs with his diary entries and news clippings. I don't know whether you'd call this part of our marketing operation, or something else, but it's part of the story that the company tells to itself and to the world.

We sometimes worry that there's too much reluctance at Brompton to talk about branding and marketing at all, at least in the normal language of those professions. It's another of those things about the company where the explanation yet again begins with the phrase: 'The thing you have to understand about Andrew …' In this case, the thing you have to understand is that Andrew has always had a dislike of anything that seemed like fashion or consumerism, and that attitude has always shaped the corporate culture. Tim Guinness was from a financial background where marketing didn't play much part in his view of a company, and I've always been suspicious of exaggerations and false promises. I feel there's an almost ethical distinction between communication and raising awareness on the one hand, and on the other the sort of thing that a large part of the advertising industry seems to regard as its job; giving people the impression that something will either solve their problems or make them beautiful, or both, with no real basis for doing so.

Andrew and I also suffered to a degree from an occupational deformation of being engineers. Andrew is more extreme in this this than I have ever been, but I sympathised – and still do – with his view that form has to follow function. It took me a while to appreciate that 'form follows function' isn't just a statement of the obvious; it's an actual philosophical view, and one that people can and do disagree with. It's a particular way of seeing the world. I look at the Brompton, or at more or less any other device or machine, and I see it in terms of what it's meant to do, whether it's good at performing that function, and how its shape contributes to or detracts from its performance. I don't really see myself in the picture, engaging with the thing; I'm thinking about the thing itself.

This can sometimes be a slightly dangerous way to look at the world, particularly if you convince yourself that it's the only valid point of view. In the early years, Andrew used to be very defensive about our marketing material. In a brochure, we would have a clear photograph of the bike, showing all the important moving parts. But Ed Donald, our first marketing manager, would have had the photograph taken in a professional studio, with lighting that showed the paint off to its best effect, or possibly with an interesting background or with a reasonably attractive person holding it or riding it. There would be proxy battles, with the real problem camouflaged behind a technical comment on some of the language used or nitpicking about grammatical errors by the copywriter, but the real point of discomfort was the idea that the bike was being presented as a fashion choice. I don't think Andrew ever really made his peace with this.

But what I've learned from marketing people over time is that a lot of the world doesn't think this way. A lot of people don't walk through a world of shapes and forms: their life is instead made up of stories and experiences. This means that if they can't see themselves using something, they won't understand how it can help them. And until they see themselves as part of a story that includes your device, there's no way of explaining the benefits it can bring.

I began to see what was going on when Ed got us featured in the fashionable lifestyle magazine *Wallpaper**. Ed was really proud of this, and he eventually got me to understand why: the Brompton had now been endorsed by a whole different group of people. Word of mouth is fine as far as it goes, but it's limited because social circles don't overlap – even in a small town, a rumour that starts at the golf club won't necessarily ever be heard by members of the radical theatre group. In much the same way, the Brompton was selling fine among people who had a certain lifestyle and view of the world, but it wasn't going to reach anyone outside that circle unless they came into contact with it.

There was little sense of urgency to tackle this when the company's biggest problem was capacity. Arguably there wouldn't have been much point in reaching out to a whole new universe of potential customers if the consequence was that the waiting time extended from fourteen weeks to twenty-eight weeks. But we were always ambitious, and always trying to expand production. It takes time to raise awareness and time for the product to take hold in new markets, so we always needed to be marketing to the consumer of two or three years in the future.

Even if that weren't the case, this sort of thinking is dangerous in the long term. You can't succeed as a bike company by not selling the bikes to people who want them. That would be a failure of our mission. Every name on the waiting list is firstly a person who wants to change their lifestyle and is being frustrated in their aim to do so; but secondly, and more prosaically, they're an opportunity for a competitor to sell them a solution that's available now.

Building a brand is an incremental process of generating awareness, explaining to consumer groups what the bike can do for them and constructing a story which allows them to begin to see themselves on the bike. Then they use the product, tell their friends and post on social media. Over the next five years, this has a compound effect on the audience for the Brompton, but only if the original customer is still happy to endorse it. That means that looking after the customer in the long term is the heart of marketing. It also means that part of the process of planning to expand output has to be, in some sense, planning how the bikes are going to be moved on to customers. It's just a bit more difficult because you can't hire customers.

What you can do, however, is to try to reach out to new groups of people who haven't heard about the bike, or, if they have, who've not yet been given a picture or story that they can see themselves in. That's the purpose of our special editions. Some have been limited edition colours, or themed releases like the Royal Wedding commemorative edition that we released in Asia. A lot of the more recent ones have been collaborations with other people and companies, people who share some of our values (and who like the bike), but who speak to an audience that we're not currently getting through to.

For example, it may sound strange, but one group of people we have historically found it hard to reach are cyclists. There are plenty of people who love riding bicycles and who are prepared to pay sums of money well in excess of even the top of the Brompton range, but they are recreational riders. We should always have been more popular with the fabled 'middle-aged man in Lycra', or MAMIL, because it seems like such an obvious extension of their hobby. As a Brompton user, I cycle about forty miles over the course of a week, going to meetings around London. I don't cycle at the weekends because I have other things to do and a young family, but if I was in a cycling club, it's hard to think that there wouldn't be a significant advantage in the weekend sportif from having put those miles in rather than sitting on an underground train.

In theory it makes sense, but in practice it was impossible to get this audience to see themselves in the picture. Until I happened to be invited to a charity event where I was seated next to David Millar. David was the young star of professional cycling in the days when there were very few British professionals – he won a Tour de France stage and the world time-trial championship. Then he was caught taking performance-enhancing drugs, confessed and was suspended from the sport for four years. But he came back clean

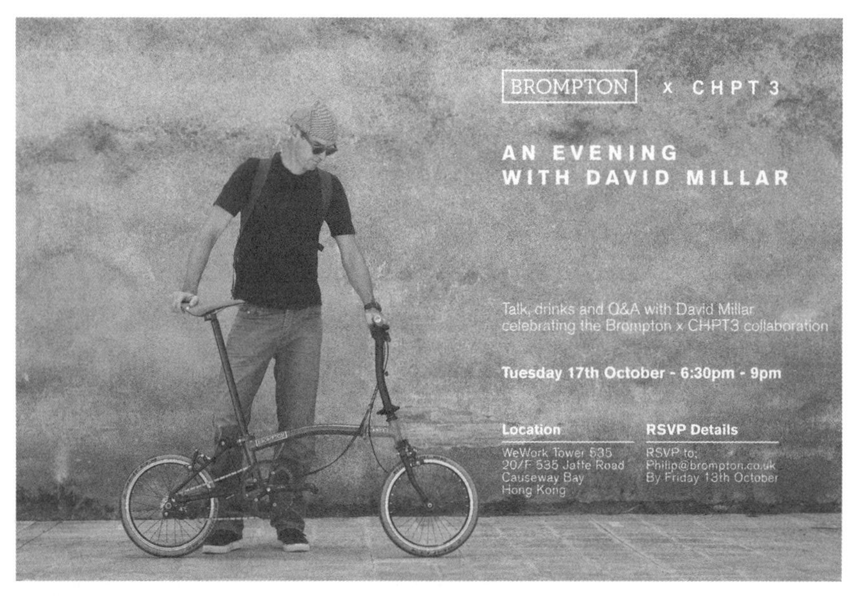

Cycling hero David Millar brought a new audience to the Brompton with the CHPT3 model.

and continued to win stages and time-trial medals while leading what turned out to be one of the few dope-free teams in cycling. It was an extraordinary story, about which he's written an award-winning autobiography, *The Racer*. He's also an extremely fashionable and stylish guy, who was launching a successful sportswear range at the time when we met.

He had tried the Brompton, because his mum had one that he used to borrow when he was in London, and he had ideas about the sort of things which would make it more appealing to bike fans. After three years, we had a model called the 'CHPT3' (the name of David's fashion label) with a titanium frame, no mudguards or luggage racks, and a lot of the super-premium components that bike enthusiasts really value – saddles, handlebar grips and slightly BMX-like tyres. We produce it as a special edition every couple of years; so far it has sold out quickly every time and the specialist cycle press have gone crazy for it.

That's something of an unusual example, because for the most part it isn't the purpose of the special editions to sell a lot of bikes. Some of them aren't intended to sell at all, like the copper-plated version we produced as a collaboration with designer Tom Dixon. That was just too labour-intensive to produce at scale, and it nearly caused a revolution in the factory. Because copper doesn't bond to steel, it first had to be plated with nickel, and the two coatings pushed everything out of tolerance, requiring vast amounts of hand finishing to make the components fit together.

But someone who wouldn't necessarily have heard about the Brompton might see the copper edition in a magazine and test ride a production model out of curiosity. And once the consumer has put himself or herself in the picture, the barrier has come down. Once they have owned it and used it, they understand the true value of the urban freedom and fun it brings. And if their limited edition is stolen after a few years and the only Brompton available is a plain green one, we find that they will buy it rather than wait, because the bike has changed their life, and the Brompton is part of them now.

What's more, talking to artists and poets – the people whose job it is to not know what can't be done – expands your horizons. The Tom Dixon copper model never went beyond a few photo shoots, but we learned about electroplating, and the nickel 'undercoat' looked amazing on its own. We've sold thousands of nickel Bromptons since.

This is what selling is about. We tell the truth about our product, to as many people as possible, in a language that they understand. That means adapting the message, looking for groups that we haven't connected to and explaining what the bike can do to change their lives. What it doesn't mean is compromising on our purpose, seeking publicity for its own sake, or exaggeration. The company philosophy that Andrew wrote in 2002 still describes our approach so well that I hardly needed to change it when I updated the marketing manual in 2016:

> Our Philosophy demands the following: To offer only products which have real practical value, to avoid developments and designs whose primary appeal is superficial and unlikely to be genuinely useful. To take great care over every detail in design and manufacturing, so that our bikes not only ride well, but can also be effortlessly folded, carried and stored. To provide well-informed, honest and accurate information about our products, and to avoid exaggerated claims for their merits. To position Brompton products as versatile tools that can enhance people's lives, not as ingenious gadgets or fashion accessories. To promote Brompton products in a non-aggressive manner and to avoid disparaging competitors' products. To look after Brompton customers, whether their bikes are old or new, and to provide service beyond what they might expect. To treat failures, mistakes and problems as opportunities to promote goodwill and build the brand.

Talking to the outside world is how you grow the brand; in order to preserve it, you have to spend a lot of time talking to your own people. One of Brompton's greatest strengths is the community of people created by the bike who have it as part of their identity and enjoy recreating the bike's personality in their everyday life. We have the Asian consumer to thank for teaching us this.

Asian customers, and particularly the Japanese, have a very different relationship with their bikes from what we were used to in Europe. It used to distress Andrew to see Japanese owners with gleaming, pristine machines dressed up with accessories; he always felt that if people were using the Brompton 'correctly' it would be beaten up, possibly chipped or dented and certainly ingrained with the dirt of a daily commute.

The point, however, is that although it's a tool, different groups of people have different problems to solve, and we needed to understand the customer. The Asian market for bicycles didn't develop in anything like the same way as the European market. In 2005, cycling was still used by many in Asia as a mode of transport and the car was very much a sign of 'success'. The cycling infrastructure was behind most of Europe, yet increasingly people were interested in the environment and their own health. And living spaces in Asia are very small, in huge cities and with a culture where family and work take absolute priority.

At the turn of the century Asian consumers were finding themselves with more disposable income and more leisure time than ever before. But there is only so much 'stuff' you can buy, particularly if you live in a very small apartment. What people were increasingly looking for was a community, a place to meet and share experiences with like-minded people. Living in a small apartment, even if you are surrounded by gorgeous clothes and handbags, can be lonely.

The Brompton unknowingly filled a gap. The product is beautiful, a work of art, appreciated by sophisticated Asian consumers; but more than that, it bought freedom, the ability to escape the city and explore. If you owned a Brompton it said something about who you were. You were smart, you could see what others couldn't, you were environmentally conscious, adventurous and interested in your health. This stimulated the formation of Brompton clubs all over Asia, and there are now hundreds of them. We didn't originally conceive this. The bike, through its fantastic flexibility, found this lifestyle, but we were flexible enough ourselves to observe this market and help it to flourish.

And in time these early adopters have gone on to lobby for cycle lanes and many have become pioneers across Asia of cycling to work. So our ultimate goal is being delivered: by a route that we would never have anticipated, the Brompton is becoming a tool for urban mobility in Asia too.

Special editions have always sold really well in Japan, partly because local riders wanted something unique that would express their personality. Word of mouth there seemed to spread not so much in workplaces as in social clubs; there were Brompton riding groups that ended up with hundreds of members. People met the loves of their lives while riding through the cherry orchards and ended up having Brompton-themed weddings. One of the signs that the bike was taking off was when it started appearing in manga comics, including some by local superstar artists like Fumihiro Katagai.

Koos Kroon, our Spanish distributor and the inspiration for the Brompton World Championships.

And Koos Kroon will forever have a place in Brompton history after he called up in 2006 from his distributorship in Barcelona and said, 'Will, I want to race these bikes.' The first two Brompton World Championships took place in Barcelona, but by 2008 the event was already too big to handle there, so we moved it to Blenheim Palace in Oxfordshire and started organising it from the factory. Since then, the

Although it's a community event, people take the Championships quite seriously.

Championships have also taken place at the motor-racing track at Goodwood, and have now settled on the Mall in the centre of London outside Buckingham Palace. We now have heats all over the world, with regional championships in the USA, Mexico, Germany, Korea and many other places, with the male and female winners then flown to London for the world event.

Part of the point of the World Championships is to establish that the Brompton can go fast. One of the biggest points of resistance people have to buying the bike is a memory of the last bike they had with small wheels, when they were twelve or so, desperately trying to spin the pedals fast enough. But a small wheel, as long as the tyre is inflated properly, has a very small point of contact with the road surface, so you lose less energy in deforming the tyre and use more of it moving forward. It also has less air resistance from the spokes. You need higher gearing to get the pedal speed to a reasonable rate, which loses you some power, but the net effect is hardly any difference. Anyone in reasonable physical condition can keep up with a cycle club on a Brompton on level ground, and they look at you as if you were a super-athlete when you do it.

But the real purpose of the World Championships is to bring people together and celebrate the community. It's meant to feel more like a massive club meeting than a promotional exercise. Other than the regional champions, competitors pay to take part and we have sponsors to deliver excellent food, including afternoon teas and gin and tonic. We have a number of rules aimed at stopping people from taking the race too seriously; you have to wear a jacket and tie, and sportswear isn't permitted. There's a 'Le Mans' style start, with all the bikes beginning the race in folded position so that efficiently unfolding it is part of the competition.

This all helps to make it more of spectacle, so as well as delivering a great day out to our customers and staff, we end up with the kind of media coverage you couldn't possibly buy. With 600 competitors, each with an average of four friends and family at the event, and another 10,000 onlookers following it online, the exposure from the World Championship is extraordinary. And it's an absolutely genuine event. It's maintaining the identity of the bike; as well as bringing the story to newcomers, a lot of the marketing purpose is just to let the people who are already aware of it have the occasional chance to look around and think 'Isn't this wonderful?'

Throughout the book I've been talking about systems that react to the world and sustain themselves, and this is the biggest system of all, extending beyond the company into the customer base, the dealer and distributor network and the wider world. It's interesting to think about what that means for us as a company; do we own the brand, or does it own us? We need to go above and beyond the promise of excellent service and reliability, to keep earning the trust of our riders. The Brompton is an expensive purchase for most people, so we want them to know that their money is being spent on an excellent, innovative product. It isn't going on big advertising campaigns or celebrity endorsements.

We will usually say yes to a good opportunity to be associated with something or someone good or interesting. These come up quite often; people seem to like being photographed riding our bikes. A Brompton appeared at the closing ceremony of the Beijing Olympics, and Prince Harry used one to get around the Invictus Games. We launched a massive crowd-funding scheme during the pandemic to provide bikes for healthcare workers to get to their jobs while the public transport system was closed down. But we don't give bikes away just for publicity. If someone has something valuable to swap for one, like the poem by Hussain Manawer that appears on our factory wall, we'll consider that, but if you give something away just because someone's an 'influencer', that devalues it for everyone else.

Prince Harry showing his father and brother how it is done, at the first Invictus Games in London in 2014.

There are certain parts of the brand that can be protected by lawyers. We have comparatively few patents, which may surprise people, but it makes sense when you think about it. The fundamental design of the folding mechanism is long since out of patent, as it was invented in 1975, and for most of the improvements we've made since then, we have preferred not to provide the world with a detailed technical description of how we make things better (and spend a large amount on patent lawyers in order to do so). Nobody else is allowed to call their bike a Brompton, or to use a similar word in a similar font. We have gone to court in the Netherlands, Belgium and Spain to protect the 'three dimensional trademark' of the overall shape of the Brompton, so that if someone else invents a different kind of folding bike, they can't make it look too like ours; this has been appealed all the way up to the European Court of Justice and may one day be ruled on there.

These are the specific pieces of intellectual property that might be identified as the Brompton brand. But if we handed those things over to another company, or sold them, then the most likely outcome is that they would be valueless within a few years. Because the truth is that the actual 'brand' of Brompton is a whole way of doing business. Without our set of values in making products, without our customer service – even without our funny little stunts and special editions – there's nothing special about an eight-letter word, a font and a certain kind of curve to the central frame. In order to preserve the value of the Brompton brand, anyone who acquired it would basically have to reproduce the entire company.

And thinking about the brand this way informs the way that we protect it. Fakes and copies like Neobike are more irritations than threats. The copies aren't any good; the people who make them don't understand the engineering and detail that deliver the integrity of the product. Neobike weren't able to copy the jigs and they subcontracted to firms that didn't understand the plans they were given. Our policy for the last twelve years has been to step hard on anything which shows up in Europe, but not to bother with trying to chase down the original people who stole the plans and tools. The European courts generally still give better protection, we have lawyers on hand, and court appearances don't involve a long-haul flight every time; it's less of a waste of time and energy than trying to fight wars in Asia. In any case, someone who buys a cheap copy is not necessarily a sale that we lost. After talking to someone at Louis Vuitton's Asian operations, I started to see the buyer of a copy as someone who wanted a Brompton, but couldn't afford one. They've got a bike that will get them around, but it won't do the job as well as a Brompton would, and every time they ride it they will know it's not the real thing. If they get more money, or decide that they actually need a better tool for their transport needs, we'll be there.

But more than that, going around suing people isn't how you protect a brand best. What the Brompton brand means is a community of people, built around a superior product that changes their lives. The way to protect that is to keep on improving the product and innovating, to keep up with a changing world and to bring new customers in. If you're in a bike race and you don't like the way that the person behind you is sitting on your back wheel, you don't stop and have a fight with them: if you do, you'll be overtaken by the rest of the field. You have to just accept that it's happening, and do your best to open up a big enough gap to make sure you stay ahead.

'the BROMPTON'

Bike of the future.

15 BROMPTON FOR EVER

The second and third Brompton bikes ever made are on display in the factory in Greenford. The first one has presumably been melted down for scrap; Andrew wasn't pleased with it, so he threw it away.

It's a fundamental principle of our philosophy of manufacturing that activity is pulled rather than pushed. You start by thinking about the ultimate goal, the thing which creates the value, and the rest of your job is planning, to make sure the resources are provided to achieve that aim, as and when they're needed. Wouldn't it be great if you could live your whole life like that?

Unfortunately, hardly any of us are sufficiently blessed to know what their ultimate purpose is, and to be able to spend their days planning and arranging systems to achieve it. Instead, we have to live going forwards in time, building up knowledge and capabilities as we can and trying to do something with them as the opportunities arise. The world wasn't made by Taiichi Ohno and the geniuses of lean production; you have to deduce what the goal is, based only on information from the past and things that have already happened. When I joined Brompton, I was just looking for something a bit more meaningful to do than manage a chemical plant for ICI. I didn't know that I'd end up with a whole new purpose in my life, but here we are.

Building a sense of purpose for a company is the ultimate task of leadership, and it becomes more and more important as the company grows bigger. As we saw earlier, it's likely that if the future of cities and transport develops in the way that it more or less *has* to, then if Brompton is going to be a part of that future we will need to change our scale even more radically. Getting from 5,000 bikes a year to 50,000 meant a factory move, a new assembly system and all the struggles described in this book. Making another factor of ten increase is going to mean doing not simply the same things, but ten times as many of them. Brompton might need multiple factories, in different places in the world, making products that haven't even been designed yet, and taking them to an entirely new group of customers through distribution channels that don't currently exist. So it makes sense to ask – what will change, and what will stay the same?

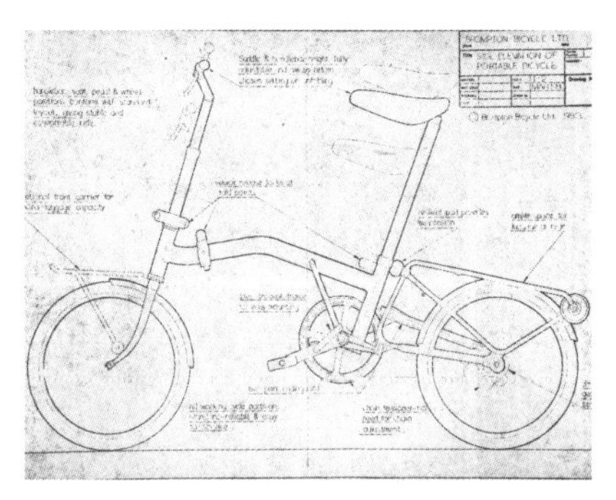

The Brompton has developed over the years, but Andrew Ritchie's original design remains at its heart.

A lot is going to change. All through this book, the point has kept recurring that growth often requires you to fundamentally reinvent your activity and do things that are not always comfortable, and it's likely that there are more such stories to come in our future. It would be suspiciously convenient if all the developments we've seen in the world over the last twenty years led precisely to the best solution being exactly the one Andrew came up with in 1975, perhaps with a few refinements to the hinge.

The Brompton bike was designed to suit the transport needs of a particular kind of urban consumer. In fact, as we've seen, it could be said that it was really designed for just one individual person, who invented it because he wanted a bicycle he could carry, and he didn't like the ride quality of the Bickerton. Not everyone is a physically fit person with enough upper-body strength to comfortably carry a bicycle, whose main need is to commute twice a day in a temperate climate, for a distance of between two and five miles, ending their journey in a place with a desk where they can conveniently store a folded package, and financially in a position where they can make a significant investment in a tool that helps to achieve this transport goal. For the last few decades, we have been concentrating on perfecting our ability to manufacture the product that perfectly addresses the needs of that specific customer, while hoping that other people can find a use for it. Starting from a clean sheet and considering what other kinds of people live in cities and need solutions for their other kinds of problems could mean something very different.

As the trumpeter Humphrey Lyttelton once said, 'If I knew what the future of jazz was, I'd be playing it now.' If Brompton had solutions for the problems of the future, they wouldn't be in the future. All we can do is learn from the recent past, and from the things that we've done over the last ten years to innovate and anticipate where we think the future will take us. But one thing seems clear – the role of the CEO is going to be very different in the future from the days when it involved visiting titanium suppliers, buying in distribution networks or changing the piecework system.

At Windsor Castle to receive my OBE from the Queen, 2015.

It was an important stage in Brompton's development when it became clearly impossible for any one individual to solve all the problems in the manufacturing process as they arose. Making the transition from a company where the job of the management was to solve problems to one where we built systems to make problems solve themselves required a lot of difficult and sometimes painful change.

It's arguable that the next big change of phase is what's happening right now at Brompton, as the company grows to a scale where it's no longer possible for the chief executive to know all the employees' names. We've built a structure and culture where we recruit the right people and then trust them to work out what to do, but it's no longer feasible for me to personally sit down and explain our principles to everyone, or to have the kind of oversight and relationship where I can give them feedback on how things are going.

It's not even possible to play a co-ordinating role as the central nervous system of the company, because there is just too much going on. During all my time at Brompton, one of my secret weapons has been a very big contacts book. At lots of important moments in the development of the company, I've been able to call on someone I've met personally and get exactly the help we needed at that time. This isn't luck; it's because I spend a lot of time networking, and I make the effort to keep those contacts current, rather than storing a phone number in an address book but forgetting the face as soon as the evening ends. In the future, there will have to be more people at Brompton with their own sets of key contacts; it might be one of the last bottlenecks we ever eliminate, but this too is going to need to scale up.

I need to make all the different teams understand that they have to communicate with each other and build outside relationships of their own. They can't just report everything to the CEO's office and assume that all the information is being collated into a comprehensive plan. This is the last thing that you have to let go of as the system grows; not only is it not possible to micromanage everything to fit your own preferences and priorities, but there actually isn't a single central viewpoint at all. At ten times the current scale, this

On the mezzanine at our Greenford factory.

will have to be a fact of life rather than a problem – at half a million bikes a year, there may be whole operations where the daily management is for all practical purposes completely independent of the CEO's office.

It's oddly reminiscent of the development of cities themselves. In the earliest days, you might have had a monastery or a garrison or a trading port, and everything was built around that. So the abbot, or the mayor or military governor or whoever, was in charge, and would be capable, at least in principle, of intervening and managing, since the town was made up of people and resources all serving the single central institution. At some scale, though, a city becomes big enough to develop its own identity. It's now a complex entity, and the systems which had originally been used to control and manage it are now services to it, ensuring that the various projects and priorities of the different parts are made consistent with each other and with the available resources.

Does that mean that a big company can't be managed? Clearly not. Even as systems grow beyond the ability of any one human being to hold them in their head, they start to develop an identity and personality of their own, and the creation and maintenance of that identity becomes the key function. In our context, it means that, increasingly as the company grows, my job and that of our management team is to set the culture and spend time defining what Brompton is – how the mission of providing freedom to urban people and making their lives better translates into physical actions in the world. I said earlier that components of the bike could be considered more or less important to manufacture in-house depending on their degree of 'Brompton-ness'; that's true of all sorts of other policies and processes. And the key responsibility at the top is to continuously define what 'Brompton-ness' means; to interpret and adapt the company DNA.

Giving a school talk, for the 'Inspiring the Future' charity that I helped found and supported as a trustee for ten years.

For example, one of our big challenges for the next few years is to really develop the business in the USA. If we are going to succeed in the kind of ambitious goals that the post-car future requires, there will be a day, not necessarily too far in the future, when Brompton Bicycle Limited has a US operation which sells more bikes than the whole company does today. That will mean solving logistical, manufacturing and distribution problems which are likely to be completely different from anything we've faced so far. It also means that we're going to need to find a way to talk to entirely different groups of consumers, some of whom may be people who start off with a completely different cultural and emotional relationship to different forms of transport from our current customer base. Starting from a blank sheet of paper, I don't know how that would be achieved. Even if I did, I might not be the right person to execute it, even if I could move my family across the Atlantic and spend a decade working on the project. The organisation is going to have to find a way of achieving the goal organically.

But that's not the same thing as saying that anything could happen. It's unlikely, for example, that the solution to Brompton in the USA is going to involve taking a massive venture capital investment, and spending half of it on a coast-to-coast advertising blitz. Whatever the Brompton of the future looks like, or however it's designed, you can be sure that it won't be a fashion product, with a new style coming out every season and an upgrade cycle built in to encourage people to throw them away every two or three years and buy a new one. With high confidence, it's possible to say that none of these things would work; the whole DNA of the company would be against them.

I don't like to reject ideas. That's partly for the reason given earlier – every time you say no to one idea from someone, you lose the next three or four ideas they might have had. But sometimes there's also a bigger issue. If someone has come up with something which I think is impractical, or wrong for the company, or inconsistent with the brand, then there's an important question – why did they think that was all right? If it's just a matter of picking one alternative rather than another as the best way to achieve a basically sound goal, then that's one thing. But if the suggestion itself is intrinsically wrong for the company, then that's a warning sign, because it's an indication that part of the philosophy hasn't been communicated properly. A strong corporate culture is the ultimate filter for the noise and chaos of the business world, because it means that everyone has a shared idea of what they ought to be looking at and thinking about.

The people who succeed at Brompton tend to be a little bit individual. They are the ones who form cycle clubs or used to sell sweets to their classmates. They got into scrapes when they were young and they might not have quite learned their own limitations, but by their mid-twenties they have usually developed enough common sense to be put in charge of something. They're passionate about the things they want to do, and they want to change the world. They try to give other people credit and know their own limitations, but they are prepared to look someone in the eye while talking to them. I hope that's the case, anyway.

Andrew has stepped back now from day-to-day involvement, but he is still a shareholder and still takes a close interest in everything we do. He and I meet and talk regularly: we don't see eye to eye by any means, but he is as passionately committed to Brompton as ever, and I think he knows I am too. But in a few decades' time, will someone else be in my chair, answering questions about some yet-to-be-invented technology by saying 'Well, the thing you have to understand about Will is ...'? Maybe. That's the thing about the future: if I knew where we were going, we'd be there already.

INDEX

Page references for black-and-white text illustrations are in *italics*. Colour plates are indicated by 'col. pl.'

E

F

G

W

ACKNOWLEDGMENTS

I hope this book has made it clear that the Brompton story is one created by thousands of people working together. First and foremost, thanks to Andrew Ritchie, for inventing his magic carpet and starting everything. And to all of the Brompton staff, past and present – to name anyone individually would mean leaving someone else out, so I won't, even though they all deserve it. And to our customers, all over the world, who have enjoyed our bikes in ways we could never have imagined. To everyone, in fact, who has had their life improved by a Brompton.
Will Butler-Adams

My thanks to everyone at the Brompton factory who helped and spoke to me, and to Andrew Franklin at Profile, who gave so many helpful comments about how to find the stories in a sea of transcripts. Also to Paul Forty at Profile, who coaxed the book through the subsequent stages. And I'd like to single out one particular Brompton customer; my mother, Hilary Davies, who was much more enthusiastic about this than any of my other projects, which really did keep me going when things looked like hard work.
Dan Davies

IMAGE CREDITS

Images reproduced by permission of:
Will Butler-Adams, pp. 66, 68, 76, 90, 91, 95, 96, 132, 161, 165, 177, 184, 185, 190, 198, 204, 263 & 266, and colour pp. 12 (middle, bottom left, bottom right) & 15 (bottom right); Sara Cox, colour p. 13 (top right); Tim Eavis, colour p. 13 (bottom left); Tony Huang, colour p. 13 (centre left); Berenda Lim, colour p. 13 (centre right); Stella Pictures/Daniel Hambury, p. 18; Jeff Tan, colour p. 13 (top left); Serena White, colour p. 13 (bottom right).

Endpaper illustration and all other images reproduced by permission of Brompton Bicycle Ltd, with these additional credits: Anna Batchelor, pp. 113, 145, 162, 213 & 255, and colour pp. 1 (bottom), 3 (top) & 15 (middle); Lara Clarke-Wardle, pp. 32, 34, 39, 55, 57, 72, 75, 79, 82, 84, 102–3, 146–7, 154, 168, 196, 206 & 228, and colour pp. 2 (top), 8 (bottom), 9 (bottom), 10 (top and bottom), 11 (top, middle and bottom) & 12 (top); Peter Hughes, pp. 62 (top and bottom) & 86, and colour pp. 7 (top) & 16 (bottom); Andy Jones, colour pp. 14 (bottom left x 2) & 15 (top); Brian Jones, p. 246; George Marshall, pp. 244 & 250, and colour pp. 7 (middle) & 16 (top and middle); Olivier Pascauld, pp. 149 & 153; Simon Sorted, colour p. 1 (top).

ABOUT THE AUTHORS

Will Butler-Adams is a chartered engineer and CEO of Brompton Bicycle Limited. He was appointed OBE in the 2015 New Year Honours, has been featured in multiple publications including the *Financial Times* and has given talks for Google and PwC.

brompton.com | will_brompton | will_butler_adams

Dan Davies is a journalist and economist. He has worked as an analyst for a number of investment banks and has written for the *Financial Times* and *The New Yorker*.

dsquareddigest

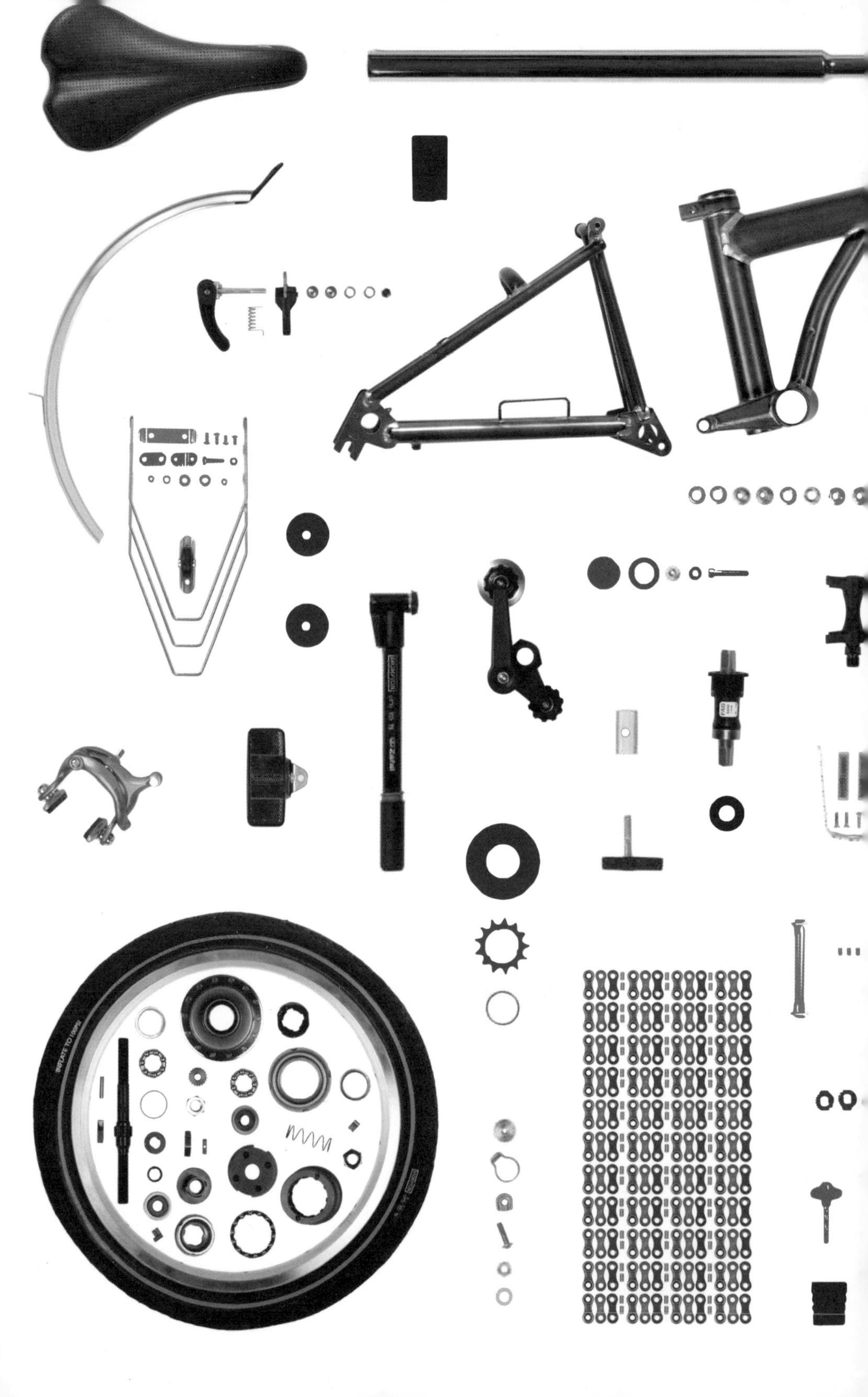
INFLATE TO 100PSI